Basic
Appliance Repair

**DISHWASHERS,
GARBAGE DISPOSERS,
ELECTRIC RANGES AND OVENS**

Basic
Appliance Repair

DISHWASHERS,
GARBAGE DISPOSERS,
ELECTRIC RANGES AND OVENS

CLIFF PORTER

HAYDEN BOOK COMPANY, INC., NEW YORK

*To my wife Virginia, daughter Mary, and son Tom,
whose help and patience made this book possible.*

PREFACE

The purpose of this book is to provide the information needed to diagnose and correct appliance breakdowns as quickly as possible. The troubleshooting and repair methods are based upon the author's experience as a Service Training Specialist for a leading appliance manufacturer and over twenty years' experience in this field.

As far as practicable, appliance components are discussed in the order in which they operate. For example, the order in which dishwasher components are covered is: timer, water inlet valve, tub, limit switch, etc. The author has found that analyzing a malfunctioning appliance in the sequence in which the components operate almost always results in diagnosing the trouble in the shortest time.

While the major emphasis of this book is on troubleshooting and repair, the operating principles of each appliance are discussed in detail. A firm understanding of how a machine and its components operate is, unquestionably, necessary for effective troubleshooting and repair.

An unusually large number of illustrations is included, and much attention was given to their preparation. Specially prepared drawings, rather than photographs, are used throughout the book to depict appliances and components. It is felt that these drawings better illustrate the components being described and their relationship to the overall appliance.

A comprehensive troubleshooting chart is given for each appliance. As in the case of the treatment of the components, troubles are listed in the order in which the appliance operates.

CLIFF PORTER

CONTENTS

Chapter 1

INTRODUCTION
TO DISHWASHERS

A dishwasher is a very efficient home appliance. Considering the amount of work it does and the time it saves, it can no longer be considered a luxury appliance. Furthermore, it washes and rinses dishes more effectively than hand washing because higher water temperatures can be used. The high water temperature and the high heat produced during the dry cycle also help to sanitize the dishes.

Many people believe that a dishwasher uses a great amount of water because it runs in continuously during the wash cycle. Actually, dishwashers generally use less water than hand washing, since the tap is usually allowed to run during hand rinsing.

Dishwashers are not complicated appliances. Basically, the dishes are washed by a splash or spray action produced by either an impeller blade or wash-arm with water that is being recirculated throughout the wash cycle. Then the dishes are rinsed by a fresh supply of hot water, which is also recirculated during the rinse cycle. Finally, the dishes are dried by hot air provided by a heating unit. In most dishwashers, there is more than one wash cycle and rinse cycle to ensure effective washing and rinsing.

TYPES OF DISHWASHERS

Dishwashers may be classified broadly as either portable or permanently installed. A portable dishwasher is mounted on casters so that it may be stored anywhere and wheeled to a sink when it is to be used. Water connections are then made to a sink or tub, and the dishwasher is plugged into a properly wired electrical out-let. A permanently installed dishwasher, as its name implies, is in-stalled at a permanent location, and its water and electrical con-nections are also made permanent at the time of installation. Most permanently installed dishwashers are mounted in or under a kitchen counter.

All dishwashers, whether portable or permanently installed, are of two basic types: impeller and wash-arm.

BASIC OPERATION

The overall cycle of a dishwasher is quite similar to hand dish-washing. For example, as in hand washing, after the tub is loaded a spray is generally used to remove the heavy soil before washing. This spray in the dishwasher not only removes some of the heavier soil, but also purges the lines of cool water before the tub starts to fill. During this time the dishwasher drain is open, as would be the sink stopper. During the wash step the drain is closed and the detergent is mixed with the wash water; an automatic detergent dispenser tilts or opens to supply the washwater with the correct amount of detergent. During the rinse step clear water flushes the wash detergent from the dishes and utensils to prepare them for drying. The water is then drained from the tub and the dishes and utensils are dried. In a dishwasher, this is accomplished by a heater which not only dries the dishes quickly, but also helps to sanitize them.

Each complete dishwasher cycle, then, consists essentially of the following basic steps:

1. Water inlet (spray and purge)
2. Wash
3. Drain
4. Rinse
5. Drain
6. Dry

In many dishwashers the wash and rinse phases are performed twice to ensure satisfactory cleaning. Furthermore, on some ma-chines a selector switch may be added to increase the variety of cycles. For example, a "rinse and hold" or a "dry only" cycle are two common selections. When the "rinse only" is used, one short wash cycle removes the heavy soil from, say, the breakfast dishes and the machine then cycles off, eliminating the remainder of the wash cycles and the dry cycle. The dishes can then be stored in the machine until the lunch and dinner dishes make up a full load for the complete wash. In some machines "dry only" may be used as a plate warmer before dinner, or used to increase the drying time after a complete cycle. Actually, these are not separate cycles, they are really phases of the complete cycle which are selected by the selector switch. The number and types of selections vary from machine to machine.

DISHWASHER CYCLE

Both types of dishwashers have similar cycles. The major difference is the way in which the wash action is obtained.

Impeller Dishwashers

The major components of the impeller dishwasher are shown in Fig. 1-1. After the dishes are placed in the racks and the door is closed, the machine is turned on by manually advancing the timer switch to the start position. On some earlier dishwashers a start button was used. From this point on, all operations are controlled automatically by the timer.

Water Inlet Phase. With the timer set to the start position, the automatic water inlet valve opens to allow water to spill or spray into the tub through the inlet water line. The machine drain remains open for approximately 60 seconds to purge the lines of cold water. Many washers also include a spray rinse of the dishes during this period.

After the purge period the machine stops draining and the tub fills to the proper level for the wash period. In some machines the fill period is controlled by the timer, lasting about one minute. Other machines contain a pressure switch that either directly controls the water level or acts as a safety device to prevent overflow. The pressure switch will close the water inlet valve if the timer malfunctions and water continues to flow.

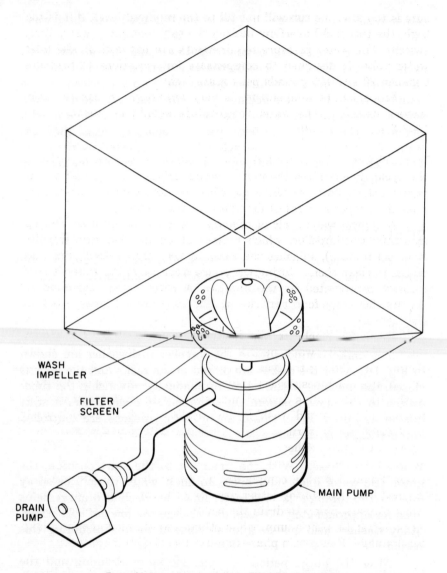

WASH
IMPELLER

FILTER
SCREEN

DRAIN
PUMP

MAIN PUMP

Fig. 1. Major components of an impeller dishwasher.

In a typical dishwasher about 8 to 10 quarts of water enter the
tub. When the time controls the fill period, adequate water pressure
is necessary so that the tub fills to the required level. If the pres-

sure is too low, the tub will not fill to the required level; if it is too high, the tub could overflow (unless the tub contains a water limit switch). The water pressure requirements are not critical; the inlet water valve is designed to compensate for variations in pressure between 20 and 100 pounds per square inch.

Proper water temperature is very important to ensure satisfactory cleaning. The water temperature should be between 140-160° F for best results.

Wash Phase. The timer has now advanced to the wash phase of the cycle. It energizes the motor, which drives the impeller or the recirculation pump. At the same time the detergent dispenser supplies the proper amount of detergent to the water.

A typical wash cycle lasts about three to six minutes. During this time the impeller, which rotates at about 1725 rpm (revolutions per minute), splashes the water at very high velocity over the dishes to clean them. During the entire wash cycle the water is continually recirculated by the impeller. A filter screen surrounding the impeller traps food particles so that they are not redeposited on the dishes.

In many machines the heater is energized (by the timer) during the entire wash period. This helps to maintain the water temperature during the wash period.

Many machines wash the dishes twice. At the end of the first wash phase the tub is drained and refilled with fresh water and detergent, and a second washing, identical to the first, is performed. In some washers a rinse phase (without detergent) is performed between the two washes.

Drain Phase. At the completion of the wash phase the water is drained from the tub by either gravity or pump action. Most dishwashers use a pump to drain the water from the tub. Some models use a separate drain pump, while others use the recirculation pump for drainage. Each drain phase requires about 45 to 90 seconds.

Rinse Phase. During the rinse phase the dishes are washed with clear water to prepare them for drying. Most dishwashers use two rinses; occasionally three rinses are performed.

The rinse phase is essentially the same as the wash phase, except that it is much shorter and no detergent is used. In some dishwashers a wetting agent is added to the water to prevent glasses from spotting. A dispenser, controlled by the timer, automatically adds the proper amount of wetting agent at the start of the final rinse.

Dry Phase.   After the final rinse, and the tub is drained, the dishes are dried by hot air. The air is warmed by the heater, which is energized by the timer. About 20 to 30 minutes are required for proper drying.

In some machines the impeller (or a separate fan) is energized during the dry phase to circulate the warm air. The impeller may operate during the entire dry phase, or for only a major part of it. Other machines use the normal convection of heated air to dry the dishes, which eliminates the need for a fan.

Wash-Arm Dishwashers

As mentioned, the operation of all dishwashers is basically the same, except for the way in which the wash action is obtained. Figure 1-2 shows the major components of a wash-arm dishwasher. Notice that they are essentially the same as for the impeller dishwasher, except that a wash-arm is used instead of an impeller. The operation of the wash-arm is much like that of a rotating lawn sprinkler in which the water pressure causes the arm to rotate. The arm rotates at a speed much slower than that of an impeller; typically about 60 rpm. The arm contains many small openings, and as it rotates water is forced through the holes at high pressure. This results in a better water pattern and more effective cleaning than can be obtained with an impeller. Wash-arm dishwashers are also generally less noisy than impeller dishwashers. Actually, the impeller principle is still used, but in a different way. An impeller blade is enclosed in a housing under the wash-arm, and is used as a pump to create the pressure needed to force the water up through the wash-arm.

A variation of the wash-arm dishwasher is shown in Fig. 1-3. An extension tube, with a spray nozzle at the top, is incorporated into, or coupled with, the wash-arm assembly. This extension ensures that an adequate spray is delivered to the upper rack of the machine even if the lower rack is overstacked or has a poor arrangement of dishes.

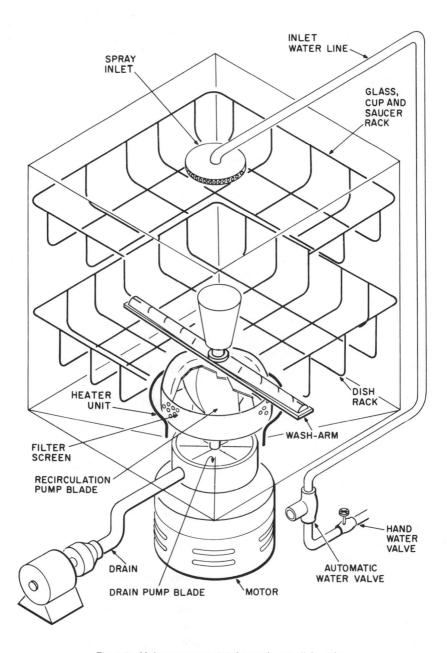

Fig. 1-2. Major components of a wash-arm dishwasher.

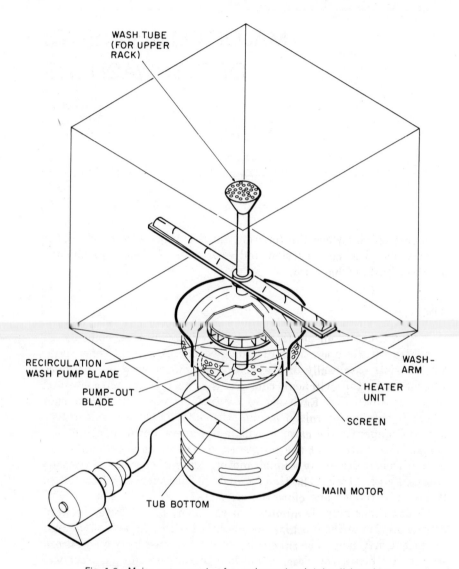

Fig. 1-3. Major components of a wash-arm/wash-tube dishwasher.

Chapter 2

MAJOR COMPONENTS
OF DISHWASHERS

This chapter describes the functions of the major components of a dishwasher. The construction and operation of these components are described in Chapter 3.

TIMER

The timer is an electrically operated device that controls all of the automatic operations of a dishwasher. A typical timer, Fig. 2-1, consists essentially of a synchronous motor with one or more sets of contacts (switches), cams, and a gear train, all contained in a single housing. Each set of contacts is controlled by a cam which is notched or raised around the outer edge. As the contact arm rests against the cam, it will move up or down, opening and closing the contacts at the proper time. The cams are mounted on a shaft that is driven by a small motor. The motor speed is reduced through a gear train. The timer moves in increments of from 30 to 90 seconds to open and close the contacts.

The timer may be mounted in the door, in the lower section, or side section of the machine, and is fitted with a knob to start and operate it manually. The timer may also be started by a pushbutton switch. If this is the case, a mechanical pushrod or an electrical holding coil keeps the timer motor operating until the starting contact takes over.

A wiring harness is used to connect the timer to each component. Each wire in the harness has a connector at one end to attach it to the timer. The connector is generally a slip fit and is referred to as a bullet or spade connector.

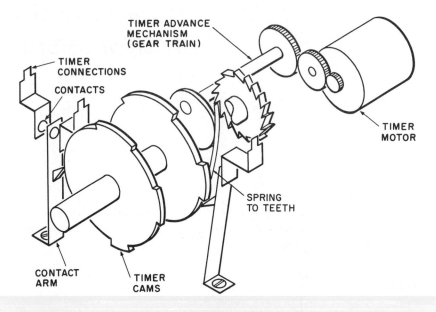

Fig. 2-1. Typical dishwasher timer.

WATER INLET VALVE

For a machine to operate automatically, the water inlet valve must be controlled electrically, Fig. 2-2. The valve consists basically of a solenoid, needle, and valve seat. Normally the valve is closed and no water can enter the tub. However, when the machine enters any phase that requires water, the solenoid is energized to move the valve needle and allow water to enter the tub.

In permanently installed machines the valve body is fitted with a plumbing connection to connect it to the existing plumbing line. For portable machines a flexible rubber hose on the machine is connected to the water tap on a sink or tub.

As the water passes through a strainer in the valve, through the valve seat and flow control washer, it enters the inlet system. The inlet system consists of either a copper or rubber line attached to a fitting at the side or on the top of the tub. An air gap which eliminates the possibility of wash water contaminating the fresh water supply is required by most sanitary plumbing codes.

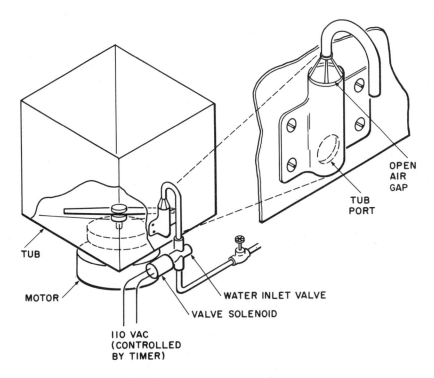

Fig. 2-2. Automatic water inlet valve.

In Fig. 2-2, a side inlet is shown; that is, the inlet water connection is installed at the side of the tub. In this arrangement, the water simply enters the side of the tub and drops to the bottom. On some machines, however, water enters at the top of the tub and falls onto a rotating deflecting wheel. The deflecting wheel sprays the water over the dishes, removing some of the soil before the wash phase begins.

DISHWASHER TUBS

The dishwasher tubs shown in Fig. 2-3 are designed for either front loading or top loading, and are generally constructed of steel coated with porcelain or plastic. The sloped well at the bottom is large enough to hold the water needed for the recirculation system to maintain the proper wash action.

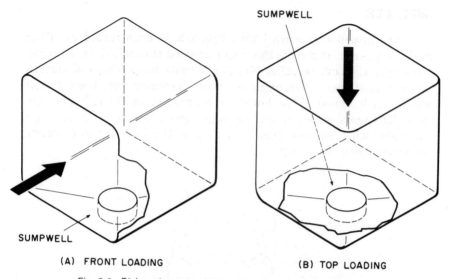

Fig. 2-3. Dishwasher tubs: (A) front loading and (B) top loading.

PRESSURE SWITCH

When the tub fills to the proper level, the pressure switch opens; this action deenergizes the inlet water valve and shuts the water off. On some machines, however, the timer controls the cutoff switch. When this is used, the machine contains a pressure limit switch, Fig. 2-4, to prevent an overflow should the timer malfunction.

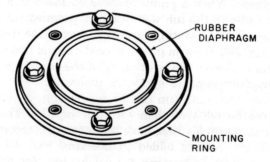

Fig. 2-4. Water limit switch used with time fill.

IMPELLER

The impeller or wash blade, Fig. 2-5, is constructed of either metal or plastic. It directs the water upward through the dish racks to remove the soil. Impeller blades generally have a curved surface and are positioned at an angle to scoop the water from the tub well and direct it upward. The lower, or leading, edge of the blade may be wider then the top, or trailing edge, and is sharp to cut through the water with less resistance and permit the impeller to circulate the water more efficiently.

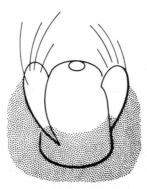

Fig. 2-5. Dishwasher impeller.

WASH-ARM

The wash-arm principle has become very popular within the past ten years. When a pump blade is enclosed in a chamber submerged in water in the tub well, water is pumped out the top of the chamber and into a rotating arm that directs the water upward as shown in Fig. 2-6. The arm pivots from the center and the force of the water, combined with the positioning of the nozzles, rotates the arm in the proper direction to spray the dishes.

Because the wash-arm can be made long, a better water distribution can be achieved than with the impeller. This is especially important in machines that have large tubs to accommodate more dishes. When a tube is added to the arm, Fig. 2-7, a dual wash action takes place, increasing the washability for the upper rack. Also, in a high tub, the flexibility for random loading is increased.

Fig. 2-6. Wash-arm.

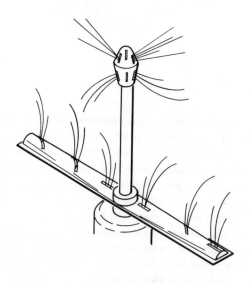

Fig. 2-7. Wash-arm with wash-tube.

WASH MOTOR

Dishwasher motors are of the split-phase, fractional-horse-power type and contain a start and run winding. The start winding, in conjunction with the run winding, is used to put the motor into motion and the start winding is taken out of the circuit when the motor comes up to speed. This is accomplished by using either a centrifugal switch governed by the motor speed, or by an electrical relay. The relay is held closed by the high current that passes through the run winding until the motor rotor (rotating part) comes up to operating speed. When operating speed is reached, the current decreases, releasing the magnetic pull on the start contacts.

DRAIN SYSTEMS

Drain systems are of two types. One type, called a gravity drain, discharges the water directly into a trap under the machine. The other type, a pump drain, discharges the water through a forced flow to a higher level before it enters the main drain system. In modern dishwashers pump drains are more common.

Pump Drain

Some pump-model dishwashers are equipped with a separate pump that is used only for discharging the water during the drain cycle, as in Fig. 2-8A. Other models combine the drain pump with the recirculating pump and operate them off the same motor, Fig. 2-8B. This is done by reversing the motor or attaching a pump to a coupling at the base of the motor. When the motor is reversible, the water is forced up through the recirculation chamber in one direction during the wash cycle, and later pumped out through the drain system when the motor is reversed.

Figure 2-8C shows an arrangement in which the drain pump is connected to the base of a nonreversible motor rather than to the top. A drain valve controls the flow to the pump. This permits the pump to operate any time that the motor is running but will discharge the water only when the drain valve is open or the overflow tube is covered.

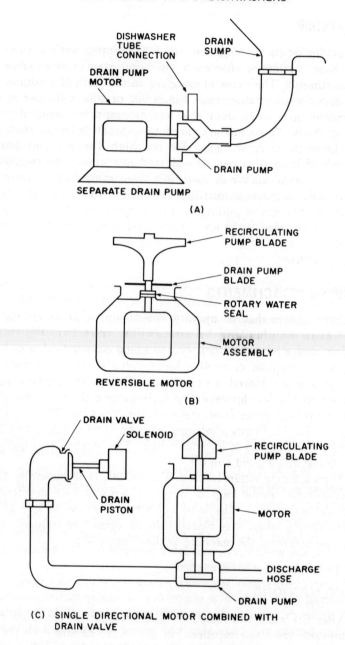

Fig. 2-8. Basic pump drain systems: (A) separate drain pump (B) reversible motor, and (C) single motor and drain.

Gravity Drain

Gravity drain systems are usually equipped with a valve assembly that is released after each wash and rinse cycle to allow the water to run out. The valve is designed in the form of a sliding piston equipped with a rubber seal, Fig. 2-9A, or with a flapper assembly resembling a small door that closes against a ground metal seat, Fig. 2-9B. These are activated by a small solenoid that controls a lever to move the seat into position. The solenoid usually has a slotted base so it can be adjusted to increase the tension on the lever. A small spring is used as a connecting link between the solenoid and the lever permitting the solenoid to seat itself without the additional strain of pulling against a rigid link. A return spring is also used to release the lever more positively when the solenoid releases.

PLUMBING ATTACHMENTS

For a dishwasher to operate effectively, it is extremely important that the plumbing be correct. The inlet line, Fig. 2-10A, is tapped from a hot water supply line using copper tubing or brass piping fitted with an accessible hand valve to act as a shut off. When the line is soldered, a union should be used to make it easier to disconnect the line for servicing or removing the machine.

The gravity drain connection at the machine, Fig. 2-10B, is usually a 1-1/2 in. fitting that may be connected to a 1-1/2 or 2 in. trap that flows into a 2 in. waste line. The waste line should be vented to permit a good run off.

When a pump drain is used, Fig. 2-10C, it is generally fitted with a rubber or copper loop to prevent the water from draining out during the wash cycle. The loop will also act as an air gap to prevent syphoning when it is placed into an open line brought up to almost the height of the machine or the counter top.

DRYING METHODS

After the wash and rinse cycles, the dishes are air dried. Some machines use the wash impeller, Fig. 2-11A, to circulate air through the machine with the lid open, or incorporate a heater, Fig. 2-11B, to raise the temperature of the air for more efficient drying. Some

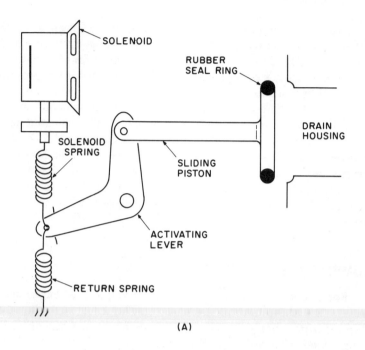

(A)

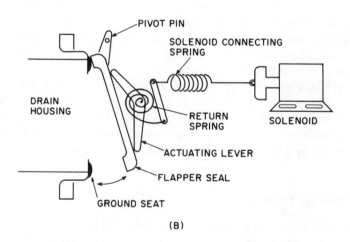

(B)

Fig. 2-9. Basic gravity drain systems

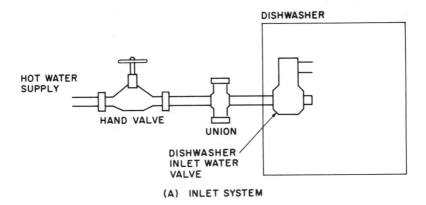

DISHWASHER

HOT WATER SUPPLY

HAND VALVE

UNION

DISHWASHER INLET WATER VALVE

(A) INLET SYSTEM

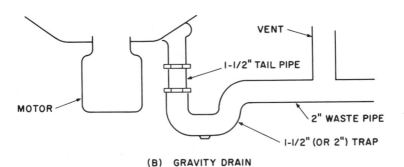

VENT

1-1/2" TAIL PIPE

MOTOR

2" WASTE PIPE

1-1/2" (OR 2") TRAP

(B) GRAVITY DRAIN

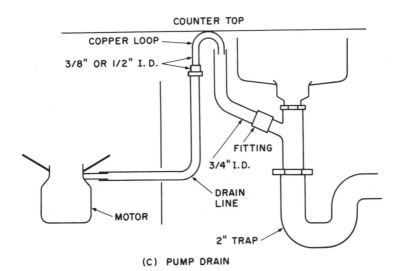

COUNTER TOP

COPPER LOOP

3/8" OR 1/2" I.D.

FITTING

3/4" I.D.

DRAIN LINE

MOTOR

2" TRAP

(C) PUMP DRAIN

Fig. 2-10. Basic plumbing system for a dishwasher: (A) inlet system, (B) gravity drain and (C) pump drain.

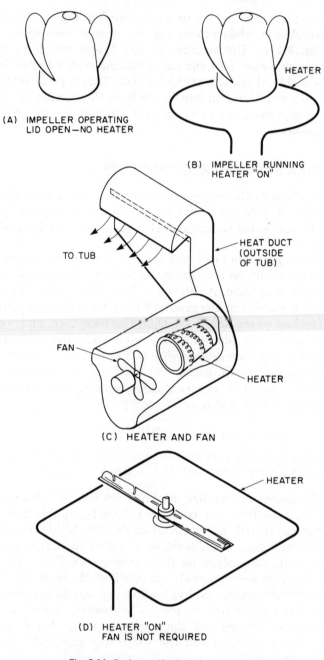

(A) IMPELLER OPERATING
LID OPEN—NO HEATER

HEATER

(B) IMPELLER RUNNING
HEATER "ON"

TO TUB

HEAT DUCT
(OUTSIDE
OF TUB)

FAN

HEATER

(C) HEATER AND FAN

HEATER

(D) HEATER "ON"
FAN IS NOT REQUIRED

Fig. 2-11. Drying methods and components.

machines use the impeller or a separate fan, Fig. 2-11C, to increase the circulation while others use the normal convection of heated air, Fig. 2-11D. The outside fan and heater will force the heated air through a duct into the tub while the sealed heater is mounted on the inside of the tub at the bottom. This type of heater may be used with or without an impeller to force the circulation. The heated air will dry the dishes and exhaust through a vent at the top of the machine.

ELECTRICAL POWER REQUIREMENTS

Both portable and permanently installed dishwashers require 110-volt, 60-cycle a-c power for operation. A portable dishwasher must be connected to a receptacle that is adequately wired to carry the load.

Most portable machines are equipped with a three-wire cord that fits a polarized receptacle. An adapter is required to fit a non-polarized receptacle. As shown in Fig. 2-12, either the small eyelet on the adapter or the ground wire attached to the adapter must be attached to the screw securing the receptacle plate, depending upon whether the adapter is used with a double or single receptacle. Make sure that the ground wire in the cord is firmly grounded at the connection block within the machine, as shown in Fig. 2-13. This is an important electrical safety requirement because the dishwasher is connected to a water supply.

NOTE

Be sure to follow the local electrical codes for the area when installing a new appliance such as dishwasher.

Permanently installed machines are not provided with a power cord. Instead they are connected directly to the fuse or circuit breaker panel with an electrical cable. The black and white wires coming from the cable should be attached to the proper terminals at the connection block on the machine, see Fig. 2-14. Whenever the washer power terminals are marked "L" and "N," the black wire, which designates the live (or line) leg, should be connected to the "L" and the white wire to the "N" or neutral terminal. When the terminals are not marked, the black wire (live) should be connected to the brass terminal and the white wire (neutral) to the nickel-plated terminal.

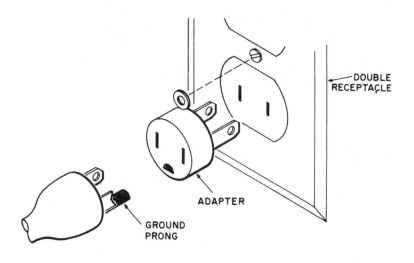

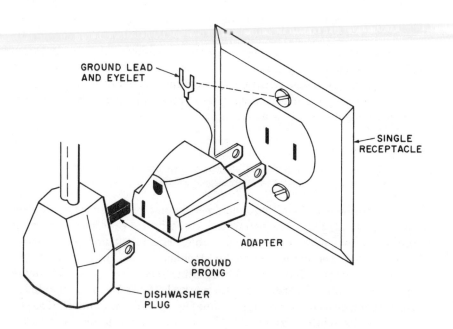

Fig. 2-12. Use of adapter with a nonpolarized receptacle.

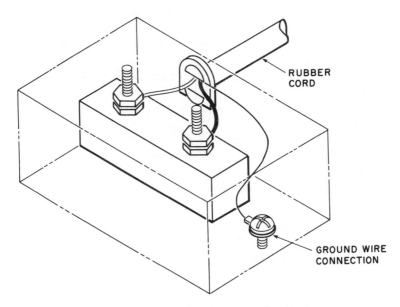

RUBBER
CORD

GROUND WIRE
CONNECTION

Fig. 2-13. Portable dishwasher connection block.

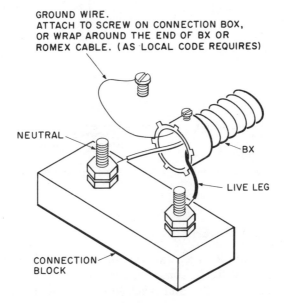

GROUND WIRE.
ATTACH TO SCREW ON CONNECTION BOX,
OR WRAP AROUND THE END OF BX OR
ROMEX CABLE. (AS LOCAL CODE REQUIRES)

NEUTRAL

BX

LIVE LEG

CONNECTION
BLOCK

Fig. 2-14. Permanently installed dishwasher connection block.

Before a dishwasher is checked out, the line voltage at the terminal block should be checked with a test lamp or voltmeter, as shown in Fig. 2-15A. If the line voltage checks all right, a further check should be made to make sure that the live and neutral wires are connected to the proper terminals. Occasionally the connections may have been crossed when the unit was installed (or previously repaired), which sometimes makes it difficult to check out the electrical circuits. The test lamp should light when connected between the live terminal and the dishwasher frame, but should not light when connected between the neutral terminal and the frame, Fig. 2-15B. If the opposite occurs, the wires on the connection blocks should be reversed.

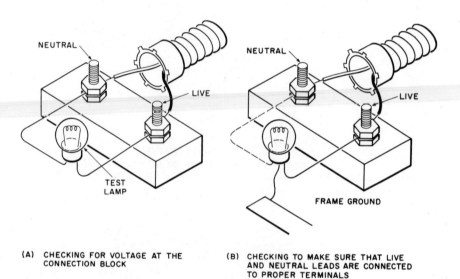

(A) CHECKING FOR VOLTAGE AT THE CONNECTION BLOCK

(B) CHECKING TO MAKE SURE THAT LIVE AND NEUTRAL LEADS ARE CONNECTED TO PROPER TERMINALS

Fig. 2-15. Checking input power connections at terminal block of a permanently-installed dishwasher.

DETERGENT DISPENSER

Dispensing the detergent at the correct time is an important part of the wash cycle. In some early model machines, the detergent cup was spring loaded and released by a cable attached to the drain lever. When the drain closed the detergent was dispensed.

Some later models either had an open cup or the correct quantity of detergent was simply placed on the open door. However, all of the detergent was used in the first wash cycle, with none left for the second wash cycle in those machines that have a double wash. Many of the later models use a tilting cup, Fig. 2-16, or a dual dispenser, Figs. 2-17—2-18, to supply each cycle with an equal amount of detergent. This is especially important in the larger machines that have a greater capacity.

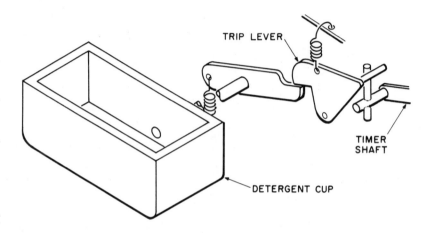

Fig. 2-16. Tilting cup.

A tilting tank may be used to act as a door for the second cup, Fig. 2-17. The cup is divided into two compartments with a small hole drilled in the inner one. During the first wash cycle both cups are filled with water and remain that way until the drain cycle when the cup with the small hole will empty. This will make the tank heavier on one side and cause it to tilt because it pivots on a shaft. A small screen covers the drain hole to prevent food particles from blocking the opening.

Two additional methods are used in the dual dispenser. One employs a tilting cup that holds the detergent, which is triggered by the timer when it trips an arm attached to the cup. The other uses a sliding door that is released by a solenoid, Fig. 2-18A, or bimetal strip, Fig. 2-18B. The difference between the two is the way they are connected electrically. For example the solenoid has the live

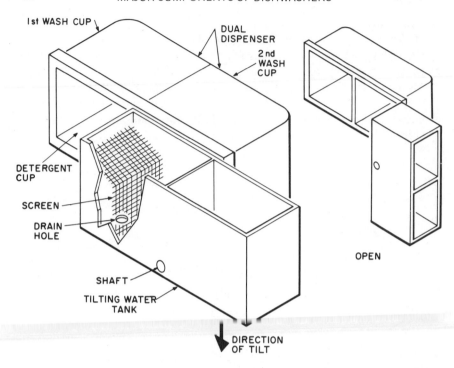

Fig. 2-17. Tilting tank.

110-volt leg and a ground wire attached directly to it. This is the same as any conventional hookup, such as a lamp socket. The bimetal has one side attached to the live wire while the other is connected to one side of the motor or heater unit whose other side is connected to ground. This will place the bimetal in series with the motor or heater and pass the current through it as it would for a piece of wire. The difference is that the bimetal will get hot and bend, thereby releasing the detergent cup cover.

RINSE ADDITIVES

To overcome the problem of glasses becoming spotted, which is caused by hard water, a rinse additive is used. A small quantity of additive is introduced into the last rinse to break the surface tension of the water and let it run off in a smooth sheet.

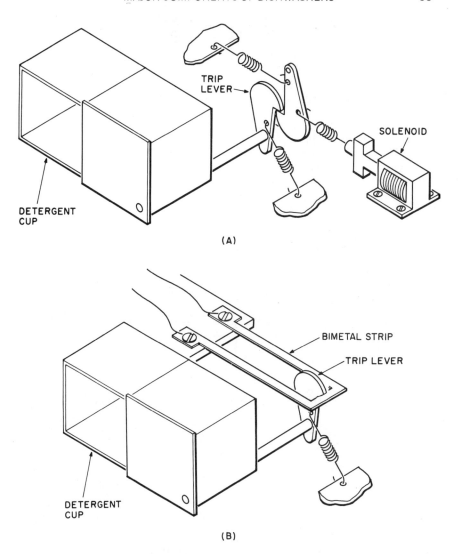

Fig. 2-18. Solenoid and bimetallic strip tilting cups.

The rinse additive injector, Fig. 2-19, may be activated by a solenoid that operates a piston which in turn forces the liquid out. When the solenoid is activated, the lever attached to the piston shaft forces it down, which sets a small amount of the liquid into

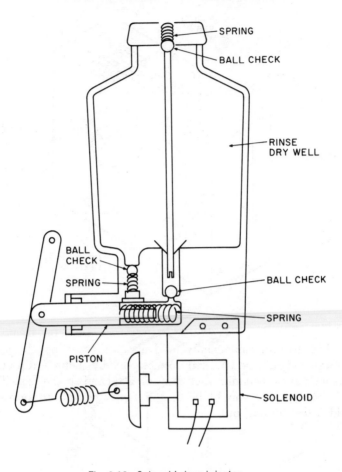

Fig. 2-19. Solenoid rinse injector.

the machine. Ball checks are used to close off the port to the tank when the piston is forcing the liquid out.

A bimetal strip, Fig. 2-20, may be used to raise a shaft that is seated against a small opening which will also allow the additive to enter the machine. When the door of the dishwasher is opened, the scoop is immersed in the rinse solution, and fills the discharge chamber when the door is closed. Current passes through the bimetal strip, which is wired in series with the heater unit or the motor winding. This bends the bimetal upward, raising the shaft and needle seat to empty the chamber of rinse liquid.

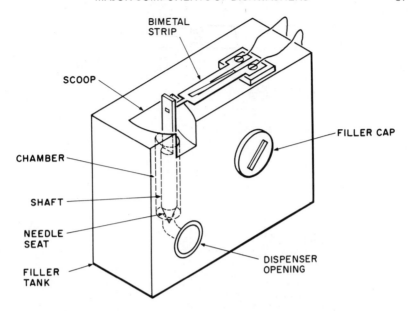

Fig. 2-20. Rinse solution injector bimetal control.

The injector can usually hold about a three-month supply of liquid and should be checked occasionally by using a dip stick. In some machines, this dip stick may be attached to the cap. The cap has a gasket and should be seated tight enough to prevent the liquid from running out.

Chapter 3

DISHWASHER TROUBLESHOOTING AND MAINTENANCE

This chapter provides the information necessary to isolate and correct most dishwasher malfunctions. First, general troubleshooting methods are described, and a dishwasher troubleshooting chart is given. Then, operation and maintenance of the major dishwasher components are discussed. As far as is practicable, these components are discussed in the order in which they operate when the dishwasher is turned on, i.e., timer, water inlet system, etc.

TROUBLESHOOTING

To check a dishwasher thoroughly, it is usually necessary to let it cycle automatically rather than rotate the timer manually. The correct water temperature is necessary during the wash cycle to do a satisfactory wash job and also during the rinse cycle to help dry the dishes. When the water is heated by a tankless heater it may run very hot for a short period and then cool off for the remainder of the cycle. Reducing the flow rate through the heater or increasing the water temperature will allow the water to have more heat after it passes through the heater. With a thermocouple submerged in the well, a water temperature reading can be taken for each fill cycle. To avoid incorrect readings, care should be taken not to clip the thermocouple to the heater, which might be energized during the wash cycle of some machines and therefore show a higher reading. A water thermometer may also be used by interrupting the cycle after each fill period. If the machine is equipped with a gravity drain, it will be necessary to hold the drain valve lever closed to prevent the water from running out.

Proper water level is another very important requirement for effective wash action. Extremely low water pressure or improper control at the inlet valve or pressure switch can reduce the water supply. The amount of water required for proper wash action will vary for different model machines and the exact amount should be checked in the dishwasher service manual. For example, if the machine requires eight to ten quarts, it can be filled manually with a one-quart bottle and the proper level determined along the tub line or the heater unit. The amount of water can be checked at the beginning of the cycle and again toward the end before it is pumped out to be sure the water is not leaving the tub because of a defective drain valve or a siphoning action.

The detergent cup should also be checked to make certain it is opening at the proper time. If the cup does not open, it will result in a poor wash; if it opens too late it can deposit some detergent in the last rinse and cause an etching on the glassware which cannot be removed.

The rinse additive breaks the surface tension of the water and allows it to run off in a sheet rather than forming droplets that will leave water spots. If the filler cap on the tank is inside the machine, it should be securely tightened to prevent an excessive amount of additive from entering the tub at one time. Excessive additive can cause oversudsing and result in leaks; see Fig. 2-20.

A good check of the heating cycle can be made by rotating the timer manually through the cycle and using a voltmeter or test lamp to make sure that the heating element remains energized during the entire drying period. However, there are times when a problem does not occur in each cycle, such as a sticking solenoid or an open valve seat, which will require more time to locate.

In summary, before beginning to dismantle a dishwasher, be sure to check the following points:

1. The correct water temperature. With a hot water thermometer, check the water at the sink faucet and in the dishwasher tub. There is about a 15° difference. On pump models the water will remain in the tub when the machine is off. Gravity drains require holding the drain closed.
2. The filter screen in the water recirculation system must be clean.

3. The proper detergent. The correct amount with the necessary additives to prevent spotting should be present.
4. Enough water in the tub. Check by opening the door slowly until the machine stops. Pump models will hold the water while gravity drains must be blocked.
5. The water level should remain constant during the wash cycle.
6. The heater should be on during the entire dry cycle.

The dishwasher troubleshooting chart lists the major troubles that might occur in any dishwasher. Refer to the pages in this book indicated in the "Remedy" column for further information that may help to clarify the trouble and for corrective action.

DISHWASHER TROUBLESHOOTING CHART

Trouble	Cause	Remedy
No start	Blown fuse or tripped circuit breaker	Replace fuse or reset circuit breaker Check for cause p. 32
	Open circuit between fuse block and terminal block	Use test lamp at terminal block If lamp does not light, check connection at circuit panel p. 29
	Open door switch	Check switch lever Use jumper at timer terminals p. 44, 83
	Timer inoperative	Main timer contact from line #1 (L1) not energizing the internal circuits Adjust contact or replace timer, if necessary p. 17, 44
	Cycle switch inoperative on multicycle machines	Make sure pushbuttons are depressed all the way Check switch circuit p. 44
No water	Hand valve in plumbing shut off	Open valve p. 25, 50

Trouble	Cause	Remedy
No water (cont'd)		Check electrical circuit at solenoid p. 56
		If light lights, check water inlet system p. 50
	Open circuit	If test lamp does not light at solenoid on water valve, check water inlet contact timer p. 44
		If test lamp lights at solenoid, check solenoid p. 57
	Screen clogged in water valve	Disconnect inlet line at valve Remove and clean screen in the valve body p. 50
	Valve seat sticking	Check valve neddle and spring p. 53
		If valve seat has a brass pilot, replace or clean p. 53
Main motor will not run	Open circuit	Check with test lamp or meter at motor connection If lamp lights, check motor If lamp does not light, check connector plugs p. 67
		Check relays p. 66
		Check timer p. 44
	Motor jammed	Dismantle wash system and check recirculating blade p. 75
Will not pump	Open circuit	Use a test lamp or meter at the pump motor terminal connections If pump has a separate motor, and test lamp lights, check for a jammed pump p. 86

Trouble	Cause	Remedy
		If test lamp does not light, check timer circuit and connections p. 49
		If wash motor is used for pump drain, check for jammed pump p. 23
		Check motor reversing circuit in timer p. 87
		Check solenoid drain control p. 87
No wash	Inadequate water level	If no water, refer to "No water" in Chart
		Check water level at completion of fill cycle p. 12, 40
		Check water level near end of wash cycle to make sure water is not leaving tub during the cycle p. 25, 84
		Check shut-off valve
		Check screen p. 71
	Tub draining during wash cycle	If machine uses a drain valve, check valve seat p. 87
		If water is being siphoned from the tub, check plumbing connection p. 25
	Wash water not draining	Check drain pump p. 86
		Check drain valve p. 84, 87
		Check plumbing line p. 25, 27
	Detergent receptacle not operating	If mechanically operated, check linkage p. 80
		If electrically operated, check circuit p. 79

Trouble	Cause	Remedy
No wash (cont'd)	Low water temperature	Check temperature p. 39
	Dirty recirculating Filter screen	Clean screen p. 71
Dishes not drying	No heat	Check connection with test lamp or meter at heater terminals p. 89
		Check timer p. 44
	Low water temperature	Check temperature p. 39
		If low adjust water heater

TIMER

Most dishwashers are started by either rotating the timer knob manually or activating a lever which is linked to the timer shaft and advances it into the start position. The timer is connected to the live side of the line, see Fig. 3-1. This opens the circuit to each component when either the timer contacts or the door switch is opened.

A machine should not have the timer on the neutral side of the line (although a few older machines do). With the timer on the neutral side, it is possible to have a complete circuit even if one of the dishwasher components shorts to ground. For example, if the solenoid or motor winding shorts near the live side of the line, the house fuse (or circuit breaker) will probably blow. If either shorts near the neutral side, it will probably have enough resistance to prevent the fuse from blowing and it will operate even though the timer is in the "off" position. Should this occur, water would run continuously. The water will simply run out through a gravity drain, but will cause flooding if the machine has a pump drain. When the motor becomes grounded and is energized from the common side of the circuit, it may run continuously and require removal of the fuse to stop it. However, a damaged run winding will draw a higher current, which will overheat it and eventually the motor would burn out.

On some dishwasher models the timer circuit may be started with a pushbutton switch, shown in Fig. 3-2. This energizes a holding coil to start the timer until a second set of contacts takes over to

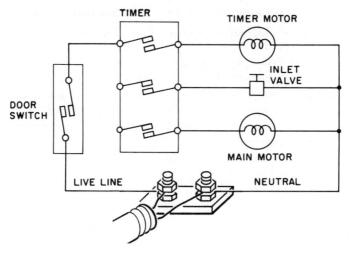

Fig. 3-1. Simplified electric circuit of a dishwasher.

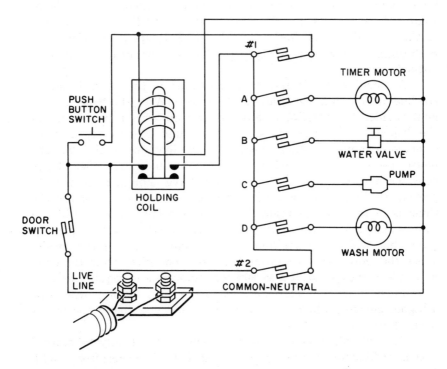

Figure 3-2. Energizing the timer with a push button switch.

keep it running through the remainder of the cycle. When the push-button switch is pressed, it energizes a holding coil, raising the plunger. This completes the circuit through contacts #1 which are closed. It also starts the timer motor, water inlet solenoid, and pump motor in sequence. After the timer advances one or two increments, contacts #2 close while contacts #1 open. This releases the holding coil and the cycle is completed through contacts #2. At the end of the cycle contacts #2 open while #1 close for the next cycle. This type of holding can be checked by energizing the top terminals with 110 volts ac and making a continuity test across the enclosed switch contacts, Fig. 3-3.

If the prongs at the top of the holding coil are spaced to fit the wall receptacle, the contacts may also be checked with an ohmmeter, see Fig. 3-4.

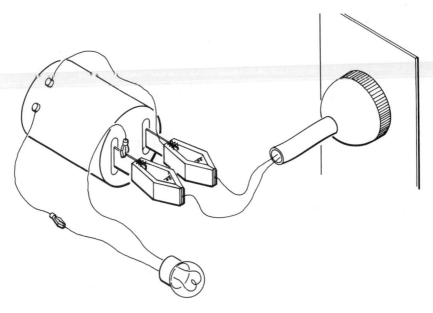

Fig. 3-3. Continuity check on holding coil.

WATER INLET CONTROL

The amount of water needed by a dishwasher can be controlled in a number of ways. First a time fill can be used, which allows the water to enter the tub for one or two timer increments, depend-

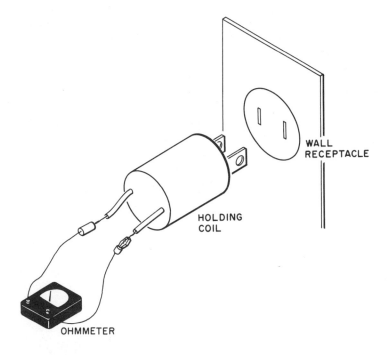

Fig. 3-4. Another method of checking holding coil continuity using an ohmmeter.

ing on the amount of water required and the time allowed for each increment. This method usually incorporates a safety switch, Fig. 3-5A, to open the circuit if the water level exceeds a maximum height. The diaphragm of the switch assembly may be mounted in the tub and is activated by the weight of the water above it. The microswitch is activated by a small pin between the diaphragm and the switch plunger arm. An adjustment nut may be used to reset the calibration for a higher or lower water level. Care must be taken not to take too many turns and exceed the limits of the switch.

Another type of pressure switch may also be used. By means of an air chamber within the pump housing or the sump well of the dishwasher, air is compressed within a plastic tube and forced against the diaphragm of a pressure switch, Fig. 3-5B. Occasionally detergents or food particles block the entrance to the tube, making the switch inactive and causing a flood.

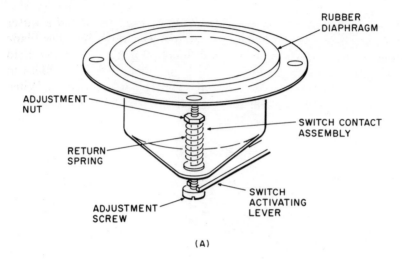

(A)

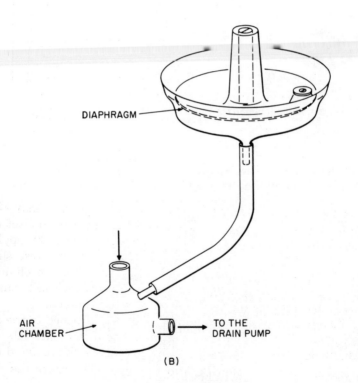

(B)

Fig. 3-5. Water level switches.

The water inlet circuit can also be controlled by means of a coil in series with the motor run winding, Fig. 3-6. As the water enters the machine the resistance against the wash impeller blade increases, raising the current and increasing the magnetic field strength of the coil. A plunger which controls a set of contacts in series with the water solenoid will open the water circuit. When the plunger within the coil is raised or lowered, the adjustment is made for the desired water level.

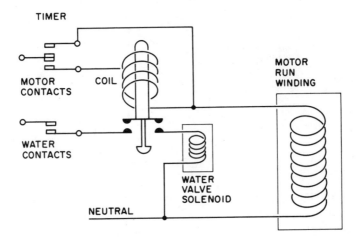

Fig. 3-6. Electrical control of water level.

Upon completion of the fill cycle the timer contacts may switch from the lower to the upper points and bypass the coil to reset it for the next fill cycle. This type of coil cannot be checked by applying 110 volts. Because of its low resistance, a continuity test should be made using a test light in series with the coil or an ohmmeter across the coil when it is deenergized. An ohmmeter check can also be made to determine the condition of the water contacts.

Finally, an overflow standpipe, Fig. 3-7, attached to the drain line, can serve as a water level control in the dishwasher tub. With a standpipe fitted to the drain casting just above the required level, any excess water overflowing into the pipe will bypass the drain valve and flow directly into the trap.

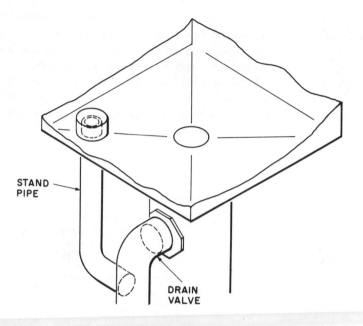

Fig. 3-7. Standpipe used for water level control.

AUTOMATIC WATER VALVES

Construction

The water supply is controlled by an electrically operated, single-port valve, Fig. 3-8, connected to the hot water line. It is important that a hand valve be installed to shut off the water if necessary. Unless the valve body can be easily disconnected from the machine, a coupling should be installed. It can become a costly job if a plumber is required to disconnect a machine for a valve body replacement. When the solenoid is energized, the valve needle is raised, allowing water to lift the valve seat. Water then flows through the center port of the valve body, through the flow control washer, and into the machine. As water enters the valve body it passes through a screen to filter out any foreign particles from reaching the valve seat.

The screen can be removed by taking off a plate or hex nut that is fitted with a copper washer to seal it. If a slow water flow is indi-

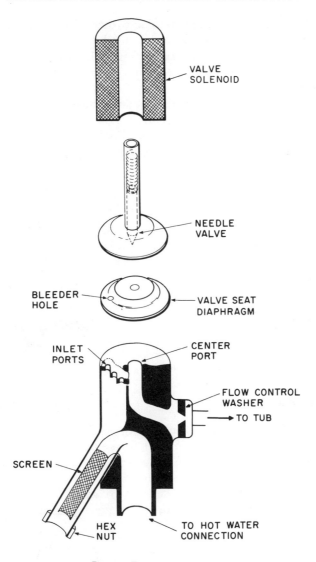

Fig. 3-8. Typical water valve.

cated it is possible that the screen is clogged and it should be checked. After passing through the screen the water enters the inlet port of the valve body.

A valve seat diaphragm is placed over the inlet ports. It is made of rubber and designed to flex as the valve needle is raised or lowered, Fig. 3-9. The diaphragm, Fig. 3-10, has a small pinhole (bleeder hole) for equalizing the water pressure when the needle is down.

The bleeder hole should be a quarter turn off the inlet ports of the valve body. This keeps the direct flow of water from entering the hole when the diaphragm is up, which could cause a chatter because of increased pressure above the diaphragm.

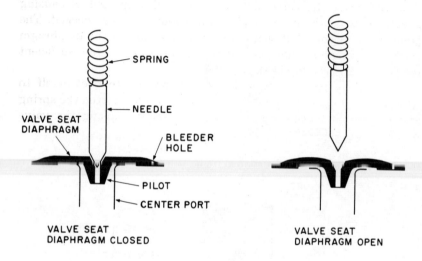

Fig. 3-9. Valve seat action.

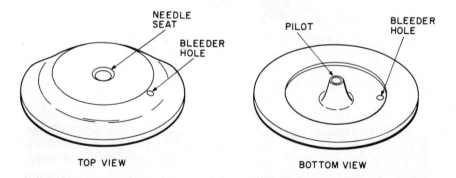

Fig. 3-10. Typical valve seat diaphragm.

A pilot guide is fitted to the underside of the diaphragm to keep it aligned in the center port. Sometimes, the pilot is made of metal. Because of close tolerances in the center port, a corroded metal guide may become enlarged and prevent the seat from either opening or closing. Removal of this corrosion with a mild cleaner sometimes makes it possible to save the seat. The metal pilot may also become loose and cause the seat to jam in an open position, causing the water to run continuously. In this case, it is advisable to replace the seat.

Whenever the valve seat diaphragm is suspected of causing a leaking valve, the needle and spring should also be checked. The point of the needle that seats into the center hole of the diaphragm should be in good condition and the spring should exert sufficient tension against the top of the needle.

Occasionally, it is possible for the end coil to hook itself to another coil, Fig. 3-11. Should this occur, it would shorten the spring and reduce its tension on the needle, possibly causing the valve to leak.

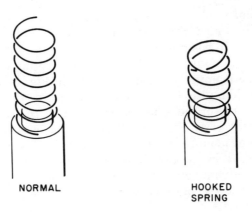

NORMAL   HOOKED
                  SPRING

Fig. 3-11. Hooked needle spring.

When assembling this type of valve, be sure that each component is properly seated, although it may be difficult when working in a confined section of the machine. For example, the valve body, which is attached to the plumbing line, may have the components facing toward the rear of the machine. It is therefore necessary to

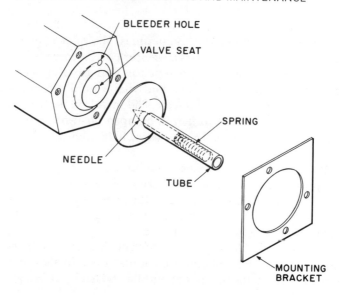

Fig. 3-12. Assembling the valve.

assemble as many of the sections as possible before joining them to the valve body, which is then attached to the plumbing. After placing the valve seat into position with the bleeder hole away from the inlet port in the valve body, bring the needle and spring (inserted in the tube) into position over the seat, as shown in Fig. 3-12.

While holding the needle tube in one hand, slip the mounting plate over the top of the tube and secure it to the top of the valve body with four screws, Fig. 3-13. When an enclosed solenoid is used it would also be mounted to the valve body. In this case, two screws hold the solenoid in place while the other two secure the mounting bracket. The advantage of this type of solenoid is that it cannot be reversed and does not require a mark to designate the "top" side.

Valve Solenoids

Three types of solenoids are used on the needle and seat type water valves, Fig. 3-14. The open coil is wound on a metal bobbin and protected with a paper or cloth covering which may be coated with baked shellac. The epoxy coil is smaller in size and has the wire coils sealed within an epoxy mold. The metal type has the coil sealed within a metal cylinder.

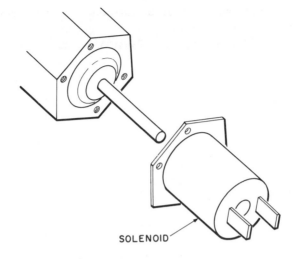

Fig. 3-13. Attaching enclosed type solenoid.

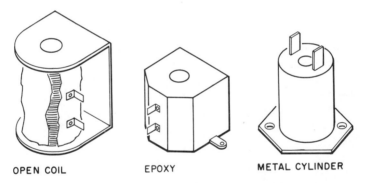

OPEN COIL   EPOXY   METAL CYLINDER

Fig. 3-14. Typical valve solenoids.

Each coil is basically the same except for the size and type of mounting. The open coil is usually held in place with a spring hooked to the rear plate or a speed nut snapped over the top of the needle tube. Care must be taken to make sure the end of the solenoid marked "top" is up toward the end of the tube so that the solenoid will pull in the the right direction. Occasionally a solenoid may be mismarked and should therefore be rechecked by reversing it if it doesn't work. Other solenoids may have tabs at the base which secure them to the valve body with screws and at the same time maintain the proper direction.

A solenoid may open up or burn out to become inoperative. A quick check at the terminals with a test light or voltmeter will assure that the circuit is closed, Fig. 3-15A. A continuity test of the sole-

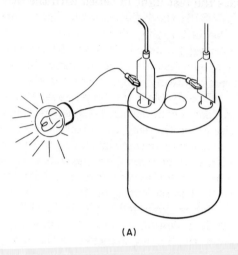

(A)

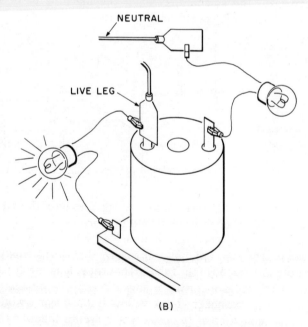

(B)

Fig. 3-15. Checking a solenoid with a test lamp.

noid coil should then be made. If the spade terminals are insulated, remove them from the solenoid and insert the test light probes to see if you get a light. If the terminals are not insulated, locate the live leg and place the test light in series with the other leg and the terminal, Fig. 3-15B. If the light does not light from the terminal to the removed connector, the solenoid is open.

When a solenoid burns out it may either open or develop a short circuit. An open solenoid will not energize, but one with a short circuit may do so by drawing a high current due to its smaller resistance. This will probably blow the house fuse. However, there is a condition which may occur either because of the manufacturer's design of the circuit or the way the machine was electrically connected. A solenoid coil may become grounded to the bobbin on which it is wound, which is in contact with the valve body and therefore the water pipe. Grounding may also occur in a portable machine with a nylon valve body. The mounting bracket is attached to the frame of the machine which is grounded through the third wire in the line cord. Usually the live side of the circuit passes through the timer contacts. Therefore, the live leg, attached to the common side of the circuit, energizes the solenoid, causing the machine to flood, even though it is in the "off" position, Fig. 3-16.

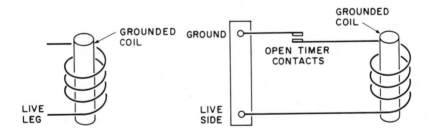

Fig. 3-16. Solenoid grounded to bobbin.

Another check for this condition is to locate the live terminal and see if it has been connected to the neutral side. If it has been, remove it. If the water continues to run into the machine the solenoid is grounded. Removal of the live wire from the solenoid, provided the connector is insulated, should stop the water. If not, the trouble is in the valve.

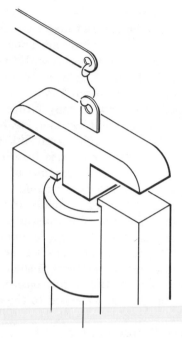

Fig. 3-17. Solenoid used for valve with external linkage.

Another type of solenoid may be used if the valve has an external mechanical linkage, Fig. 3-17. The armature, which is part of the solenoid, is attached to the valve arm with a connecting link. Occasionally the armature may not release due to residual magnetism. This may not necessarily happen in each cycle, making the trouble more difficult to locate. The armature may also release itself before it can be checked.

This type of valve may also have a small cup called a dashpot, which has a small hole in the bottom and a leather piston which moves up and down, shown in Fig. 3-18. The dashpot slows down the action of the valve arm, thereby eliminating water hammer in the pipes as the valve closes. It is important that the opening in the cup is clean in order to keep the piston from sticking. If the leather washer becomes watersoaked from a leak in the valve it should be replaced with a dry one or allowed to dry out. The metal washers in the valve seat should be placed over the rubber washers with the unfinished edge facing upward to prevent cutting into the washers, Fig. 3-19.

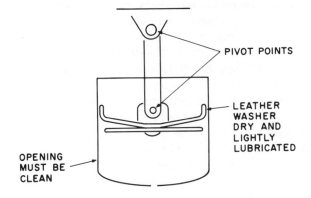

Fig. 3-18. Use of dashpot to slow valve action.

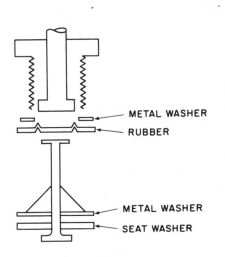

Fig. 3-19. Placement of metal and rubber washers on valve seat.

Valve Problems

The solenoid is activated by a set of contacts within the timer. At the beginning of the cycle the timer is either rotated to the start position, or started by a pushbutton switch, to energize the water solenoid for the first rinse. This is done primarily to clear the water line and furnish enough hot water for the first wash. If the water does not enter at the beginning of the cycle a series of preliminary checks should be made before dismantling the timer enclosure. If the timer were put through its cycle manually, the washer would not operate if the fuse were blown or the door switch was open. It is always wise to check the shut-off valve, which is usually either under the sink or in the basement, to make sure it is not turned off.

Within the needle and seat type valves, which are the most commonly used, a hum may be noticed when the water is turned off to the valve and the solenoid is energized. With the water off and the solenoid energized, the needle will rattle in the tube when the pressure is off. If either of these conditions does not occur, the next step would be to check the electrical circuit at the solenoid. A test light, connected across the two terminals will indicate a closed circuit if the bulb lights. If the bulb does not light, the timer should be rotated to see if the light will come on in any other part of the cycle. If it does come on, a defective cam is indicated. No light at the timer terminals indicates an open contact.

FLOW CONTROL

A flow control device is used to maintain constant flow because the water pressure may vary from 20 to over 100 psi, Fig. 3-20. The device may be a rubber washer that will flex and decrease the orifice or move against a needle point to decrease the opening as the pressure increases. The flow control must move to open or

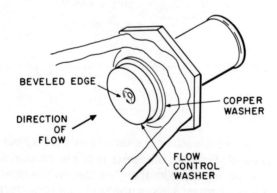

Fig. 3-20. Flow control device.

close the water port as the pressure changes. If it is a rubber washer, the rubber may harden and will not flex properly, which will raise the water level and may cause the machine to leak. Manufacturers sometimes include an additional flow washer with a larger opening for low pressure areas. If the washer has to be changed, care must be taken to install it with the orifice in the correct direction; otherwise, the opening may close off too much and restrict the flow.

Water Valve Fitting

The fitting at the water valve may be a flare, ferrule, or a rubber hose connector, depending on the type of water line used from the valve to the air gap assembly. It is important that these fittings are made up correctly because a leak at this point may not be noticed and eventually cause damage to the linoleum or flooring under the machine. In some cases the flare fitting used may be a size larger than the tube size and a reducing nut is required, Fig. 3-21. However, the flare must be big enough to cover the opening and care should be taken not to stretch the metal too much and cause it to weaken and split.

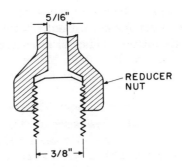

Fig. 3-21. Use of a reducing nut.

Making the flare in two steps will give the tubing more support and allow it to form properly, Fig. 3-22. Start the flare in the normal position and then raise the tubing to complete it, forming the flare gradually.

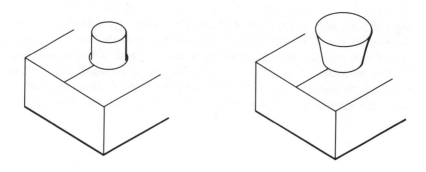

Fig. 3-22. Making a flare.

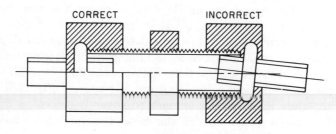

Fig. 3-23. Seating the ferrule.

If a ferrule connecter is used, the tubing should be seated all the way into the fitting. This will keep it aligned and allow the ferrule to seat properly, Fig. 3-23. When the tubing is not straight the ferrule does not lock against it properly, which may cause the fitting to leak.

When a rubber hose is used, the proper clamps should be installed to make sure the hose is secure. If a production clamp, which is pressed on with a special tool during manufacture, is used, it should be replaced with a screw type clamp unless the tool and replacement clamps are available.

Machines with copper tubing supplying water to the inlet unit may require tube repair or replacement if water in the line freezes and splits it. If the line is difficult to remove, it may be possible to cut off the damaged section and replace it with a new piece. When a piece of tubing is swadged at one end it can be soldered to the existing section and fitted to the valve with a flare or ferrule fitting, Fig. 3-24.

Fig. 3-24. Soldering an extension onto tubing.

Air Gap

The air gap, Fig. 3-25, must be designed in accordance with the local plumbing code requirements. When an air gap is required, it must be a certain distance above the water level. The gap prevents contamination of the fresh water supply. The air gap assembly is usually mounted to the side of the tub and held in place by machine screws and sealed with a gasket between the outer tub and the inlet horn. It may be difficult to hold the gasket in place when reassem-

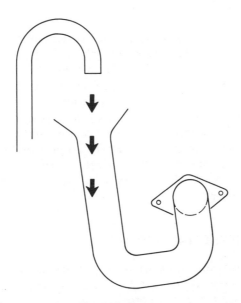

Fig. 3-25. Air gap.

bling the unit on an under the counter model. A small amount of rubber cement will keep it in place while the screws are installed and tightened. A burr or bend at the water nozzle on the gap may deflect the water to bypass the inlet horn and cause a leak. Water flowing from one line to another with an air gap in between should be directed toward the outside of the bend on a curved inlet horn. This will help to reduce the possibility of a splash out on the initial surge.

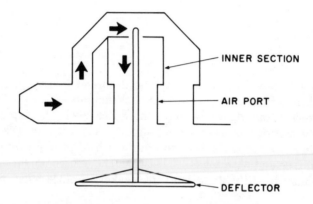

Fig. 3-26. Air gap provided by spray nozzle.

Another method of creating an air gap is to use the spray nozzle which directs the water from the top of the tub past a deflector that rotates and sprays the dishes, Fig. 3-26. This unit, which has air gaps in the inner and outer sections to act as an antisiphon, may leak if the water pressure is extremely low. Instead of the water flowing down along the deflector shaft, it may flow along the inside wall of the inner section and out the air ports. This condition may be prevalent where a well supplies the water or at the upper floors of some tall buildings. In these cases it is important that the screen in the inlet valve is clean and the flow control is operating properly.

DISHWASHER TUB AND MOTOR

Dishwashers are either front or top loading with tubs of either porcelain- or plastic-coated steel. If the porcelain becomes chipped it can be touched up with a porcelain paint to prevent the tub from rusting through. If a tub splits at a seam the area should be thoroughly cleaned and epoxy used to seal the opening. Where the opening is too wide for this, a piece of plastic or aluminum tape may be used on the back side to support the epoxy while it hardens. Ordinarily, it takes about 24 hours to harden and set properly but an infrared heat lamp can speed up the process to about 2 hours. The lamp should be positioned about six to ten inches from the work area. Be careful not to overheat any plastic sections in the tub. Where more support is needed in an area that has corroded, a piece of sheet aluminum or steel secured to the back side of the tub may then be sealed and coated with epoxy. Of course, it is necessary to remove the corrosion in and around the area with a wire brush and sandcloth to provide a firm surface for the sealer. Plastic coated tubs may also be repaired with an epoxy which is available in colors to match the tub. A manufacturer may also supply a tub bottom which may be installed and sealed with epoxy if the original tub is badly corroded and cannot be easily repaired.

On some dishwashers an insulating material may be used on the outside of the tub to help maintain constant water temperature and also reduce the sound. This is helpful if the machine is located over an uninsulated crawl space, since the cabinet section may get quite cold during the winter.

There are different tub designs to accommodate various mechanisms. For example, a tube through which a shaft passes to drive the wash impeller, Fig. 3-27A, may be part of the tub assembly. Or, a tube may be a separate part attached to a drain casting which is mounted to the base of the tub and sealed with a rubber gasket, Fig. 3-27B. Some gaskets require a sealant to ensure a good seal. Whenever a leak appears around a gasket it is important to check it carefully before dismantling the assembly. The tub may be fitted with track gaskets which are mounted to the sides and held in place with screws that pass through the tub holes. A leak at this point is difficult to see and trace against a porcelain surface and may appear as a leak at the drain casting at the bottom of the tub.

The tube must be higher than the normal water level to prevent a leak, but a water limit switch failure, which may be combined with a pump blockage, can allow the water to enter the tube and wet the motor. Some motors are equipped with a slinger, a disc slipped over the motor shaft, to direct any water away from the motor housing. However, this is only effective while the motor is running and the amount of water is not great. If the start winding of the

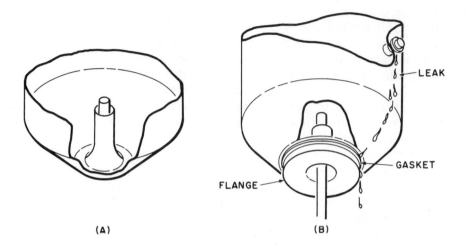

Fig. 3-27. Dishwasher tub.

motor is controlled by a centrifugal switch in the motor the water may close the gap between the contacts while the motor is running. This will energize the start winding, which is not meant to be in the circuit for any prolonged period, and cause it to burn out.

The motor will not always burn out in a flood condition; whether it does or not depends on the amount of water that entered the motor. However, tests should be made to make sure the motor has not been damaged. Of course, a motor that has been submerged in water should be completely dismantled, dried, and relubricated; but this procedure is usually not necessary when the water has come from the shaft tube. Water overflowing and collecting in the end bell should be absorbed with paper toweling. Start and stop the motor several times to help dry off the switch. The action of the switch and the circulation of air will help dispel any water droplets. An ammeter should be used on one leg of the motor line to make sure the start switch is open and the run winding is not drawing too much current. For example, a 1/4 to 1/3 horsepower motor, which draws from 3-1/2 to 5 amperes under normal operation, would indicate a damaged winding or a bound bearing if the current increases by about 3 amperes or more. This condition will eventually overheat the winding and burn it out.

A motor may be controlled by a remote relay, Fig. 3-28, located in the control or connection box. Relocation of the starting

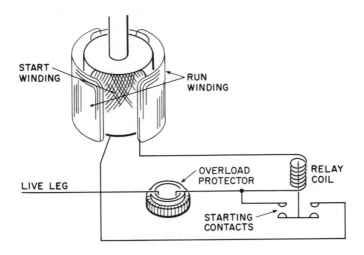

Fig. 3-28. Motor controlled by remote relay.

mechanism makes it possible to eliminate the centrifugal switch and replace the remote relay without changing the motor. This will also prevent switch failure due to flooding. However, if the water enters the bearing area it will eventually wash out the lubrication, causing the bearing to overheat and seize. A fuse or a built-in overload may be incorporated in the circuit to prevent damage to the motor field windings or open the circuit in case of a shorted or grounded winding.

If a remote relay is not used, an overload cutout is generally mounted in the end bell of the motor or attached directly to the winding to protect it against high current and overheating. If water reaches the cutout it may ground a contact to the motor frame causing it to trip. The water may also bypass it by shorting across the contact points, which would eliminate the protection.

An inoperative or open overload protector may be checked by connecting a jumper across the overload contacts to bypass the contacts, or by making a continuity test. If the jumper is used, the motor should be run for only a short period; an ammeter should be used to make certain that no more than the rated current is drawn.

A continuity test can be made with either a test lamp or an ohmmeter. If a test lamp is used, set the timer to the "on" position and connect the lamp across the overload protector contacts. A defective overload will cause the lamp to light. If an ohmmeter is used, disconnect the power source and remove a connection to one side of the overload protector. A defective overload will not deflect the needle.

Defective overload protectors of the snap-on type can easily be replaced on the motor and bell.

Another tub design, which eliminates the tub tube, uses a rotary seal to separate the motor assembly from the water compartment, Fig. 3-29. A drain well is cast as part of the upper motor housing and fitted with a valve operated by a solenoid. The drain well keeps the water in the tub during the wash cycle and discharges it into a gravity drain or pump drain when the valve opens. A small hole placed in the side of the motor housing discharges any water that passes a damaged seal, protecting the motor mechanism until the seal is replaced. For this reason the hole should never be plugged or taped.

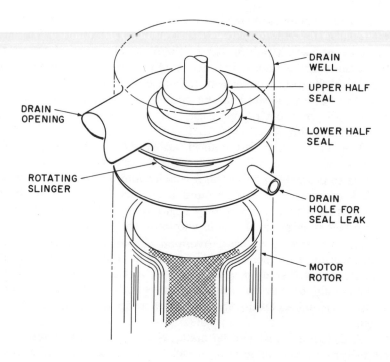

Fig. 3-29. Tub using a rotary seal.

Care should be taken when installing a new seal to make certain it is seated properly. The spring-loaded half of the assembly, which usually has a molded carbon face mounted in rubber, must be pressed into the drain casting, Fig. 3-30. The edge is coated with rubber cement which not only lubricates, making it easier to install, but also seals the rubber in the metal casting. The upper half of the seal is either a bronze or steel disc with a polished surface which rotates on the stationary section. A coating of light oil between the seal surfaces ensures smooth operation of the seal. The rotating half of the seal must also be sealed to the driving shaft if the shaft passes through the center of it. Washers are required to prevent any leakage at this point.

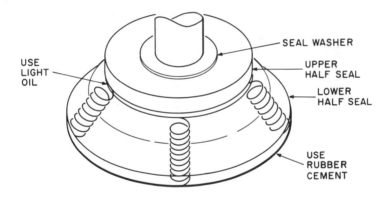

Fig. 3-30. Installing a rotary seal.

When an impeller blade is screwed onto a threaded motor shaft, the seal face may be molded into the back of the blade, Fig. 3-31. If the motor shaft ends within the blade and does not pass through it, the impeller acts as a shaft seal. Proper clearance is required between the end plate of the pump and the blade to avoid rubbing and also to maintain the correct spring pressure against the seal. Rotary seals that are allowed to run dry for any length of time may become damaged from overheating as they are kept cool by circulating water.

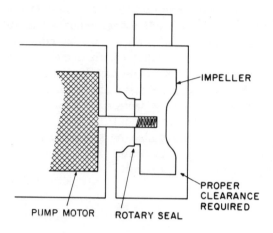

Fig. 3-31. Rotary seal molded to impeller blade.

IMPELLER

The wash blade, or impeller, which is coupled to the motor shaft, scoops up the water in the tub well and directs it toward the dishes. The motor, with speeds generally of 1725 rpm, supplies the force required to do the washing, and the contour of the blade directs the water upward to form the wash pattern. For better wash action, a deflector may be used in the lower basket to control the spray. Although the principle is simple, the components must be in good condition and operate properly to get the proper results. If the blade is chipped or bent, it may not pick up enough water, which would cause a poor wash, or it may cause a leak to develop by directing the water toward the air vents used for the drying cycle. A badly damaged blade can develop a rumble because it is unbalanced and can cause excessive wear on the upper motor bearing.

The impeller is held in place by a long bolt passed through the impeller shaft and threaded into the motor shaft or an adapter fitted to it, shown in Fig. 3-32. Although the bolt is made of brass or stainless steel a gasket is often needed to prevent water from running down the bolt and causing the threads in the steel motor shaft to rust. Care should be taken when removing the bolt, since too much leverage with a long wrench may twist the top off. Once the bolt is started, a little rust cutting oil applied under the head will flow down to the threads and help to loosen it. A new gasket should then be installed with the new impeller to keep the water out.

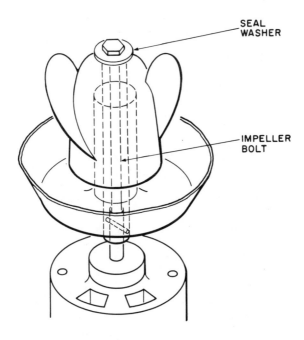

SEAL WASHER

IMPELLER BOLT

Fig. 3-32. Impeller bolted to motor shaft.

The impeller shaft bolt on rotary seal models, Fig. 3-33, not only couples the impeller to the motor shaft but maintains the required pressure on the seal to keep it water tight. If water is seen' coming from the weep hole at the side of the motor, an impeller bolt may be loose.

A filtering screen prevents food particles from being recirculated with the wash water, Fig. 3-34. It is important that the screen be cleaned periodically for good wash results. If enough water is not filtered through the screen to the impeller blade, the wash action will be reduced.

A protective screen, Fig. 3-35, may be installed in the bottom of the basket or covering the impeller to prevent a dislodged utensil from damaging the blade during the wash cycle.

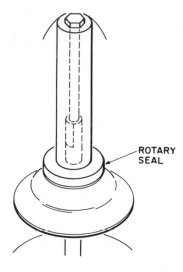

Fig. 3-33. Impeller bolted to motor through rotary seal.

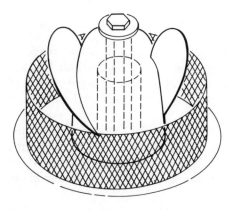

Fig. 3-34. Filter screen.

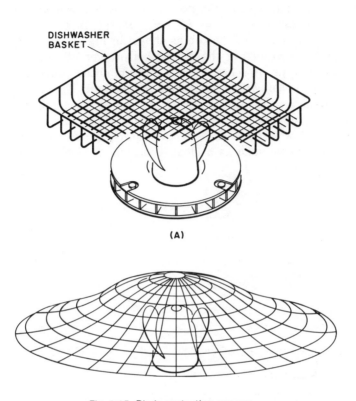

Fig. 3-35. Blade protection screens.

WASH-ARM

The wash-arm is rotated by the water being pumped into it. In a wash-arm machine, an impeller blade is enclosed in a housing where it functions as a pump blade to develop the water pressure in the wash-arm. Water is directed through the housing by baffles or vanes to the wash-arm, which is pivoted on a pin and rotated by the water's force. The arm is equipped with several slots to rotate the arm in the proper direction and direct the water toward the dishes in an effective wash pattern.

The blade may be driven at 1725 or 3450 rpm, depending on the motor speed. The impeller pump blade must be below the water level to maintain a constant flow to rotate the arm at the proper speed, approximately 50 to 60 rpm. If the motor is vertically mounted to the base of the tub as shown in Fig. 3-36, the recirculating pump is installed in the tub well. However, the pump may be an external type attached to a motor mounted horizontally and coupled to the tub with rubber hose connections, Fig. 3-37.

For the recirculating pump to operate at maximum efficiency, it must be supplied with sufficient water. This supply can be affected by a low water level from low water pressure or a restriction at the inlet valve such as a clogged screen. A defective or an im-

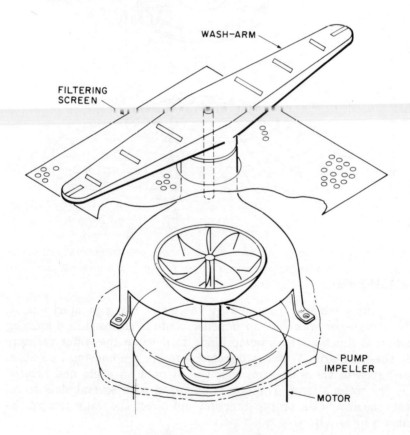

WASH–ARM

FILTERING SCREEN

PUMP IMPELLER

MOTOR

Fig. 3-36. Wash-arm mechanism.

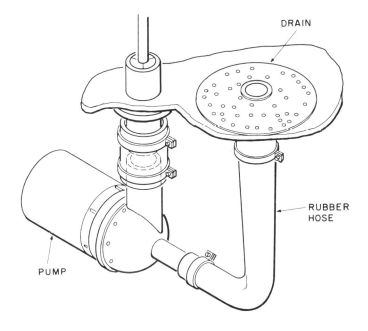

Fig. 3-37. External pump.

properly adjusted pressure switch may also limit the water required in the machine. Although these are often the probable causes, it is most important to check the filter screen in the recirculating system before dismantling or adjusting any of the other components. The screen may be self-cleaning, being washed down during the drain cycle, and may act as a filter during the wash cycle. This is done by positioning the recirculating pump blade above the filter screen and placing the drain opening below it, Fig. 3-38.

If the water flow becomes restricted when an external pump is used the recirculating hose may collapse and shut off the water supply completely. The water supply can also be shut off completely, or greatly restricted, if the hose is twisted during installation.

The pump blade should be in good condition and securely tightened with the proper seal washers to keep the water from reaching the motor shaft and entering the upper motor bearing. If the motor is designed to operate in either direction, a lock for the impeller bolt should be used. A lock washer or a nylon insert in the threaded part of the bolt will prevent it from backing out.

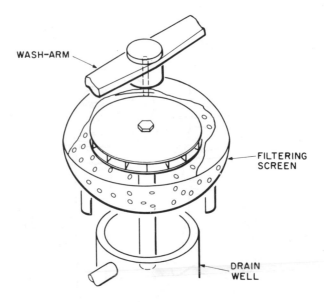

WASH-ARM

FILTERING
SCREEN

DRAIN
WELL

Fig. 3-38. Self-cleaning screen.

Deflectors and water separators are often incorporated into the pump housing to direct the water to the opening and prevent recirculation within the housing. The deflector, Fig. 3-39, may be notched to fit a key in the housing to maintain its proper position, since improper location of the deflector may block the water flow to the separators and adversely affect the wash action.

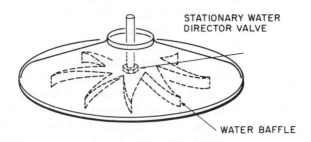

STATIONARY WATER
DIRECTOR VALVE

WATER BAFFLE

Fig. 3-39. Deflector and separators.

The force of the water leaving the pump will be directed into the wash-arm, causing it to rotate and to spray the dishes. The openings in the arm are positioned at the proper angle to drive it and direct the water in an effective wash pattern. It is important that the arm rotate freely to do its work. Some models have a single arm, which is held in place by a lock ring on a pivot pin or by the weight of the arm itself. If the arm is made of metal it may be necessary to lubricate the pivot pin periodically to prevent binding. When the arm is made up of upper and lower sections, the sections must be held together firmly to avoid a water spray from the joint. This could direct the water toward the base of the door or toward an air vent in the lower part of the tub, causing a pulsating leak. The pulsating leak would take place each time the open section of the wash-arm passed the door or vent opening.

In addition to the single wash-arm, a tube or extension may be coupled to the arm to increase the wash action for the upper rack. As shown in Fig. 3-40, the extension may be a telescoping tube raised by the water pressure in the arm, or a tube coupled to the wash-arm by a clutch, basically a rubber disc that engages a flange at the base of the tube and drives it at the same speed as the main wash-arm. The permanently coupled and telescoping wash-arm extensions are attached to the wash-arm and require a slot or opening in the lower rack.

The clutch type rotates in a bearing brace and is mounted to a support assembly in the rack. It is important that the tube rotate freely and that proper clearance is maintained between the wash-arm clutch and the tube flange. Proper clutch clearance depends on the design. For example, the expanding clutch requires approximately 3/16 inch and the raisable wash-arm may be adjusted from 1/4 to 3/8 inch, as shown in Fig. 3-41.

The expending clutch should be in good condition and properly aligned with the upper tube flange. If the opening in the rubber clutch does not meet the opening in the tube, the proper coupling action may not take place and the water may be forced out at the flange instead of going up through the tube for the upper wash action. This may cause a tilting type detergent cup on the door to flip and use all the detergent in the first wash. A split in the expanding clutch may also cause a pulsating leak at the bottom of the door each time the wash-arm comes around to the front of the machine.

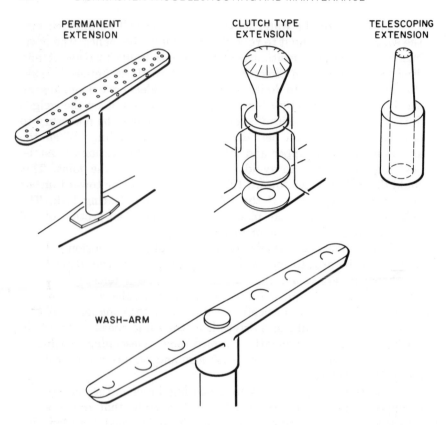

Fig. 3-40. Wash-arm extensions.

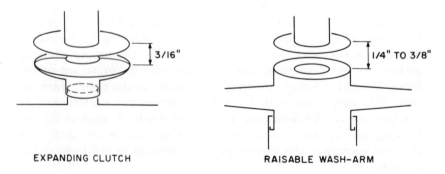

Fig. 3-41. Clutch clearance.

Another method that uses the pump type wash action has a hose that brings the water to a tube inside the wash tub, Fig. 3-42A. The tube is rotated by the water action and develops a circular wash pattern. The tube may be placed horizontally in the center of the tub and supply the spray for the upper and lower racks. A hose may also be used to supply a wash-arm or disc mounted to the top of the tub to increase the wash action, shown in Fig. 3-42B.

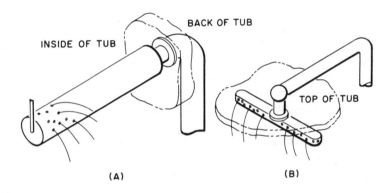

Fig. 3-42. Other methods of obtaining wash action.

DETERGENT DISPENSERS

For good dishwashing results, the correct type of detergent in the proper quantity and at the right time is very important. Depending on the dishwasher model, the detergent may be placed on the door or in a cup to mix with the water in the wash cycle. Some models have two wash cycles and have a double cup to meter the detergent for each cycle.

A covered tilting cup may be used, which is triggered by a cam in the timer or energized by a bimetal strip. The cover prevents the detergent from washing out until it is needed for the wash cycle. When the cup is set in the fill position and the timer is rotated manually, the cup should tip at the beginning of the wash cycle.

The bimetal type has a metal strip which heats and flexes from the current passing through it. Its resistance is very low and it must be in series with one of the components, such as the heater or the motor, to prevent an excessively high current from damaging it, Fig. 3-43. Never apply 110 volts directly across the bimetal, since this would cause permanent damage. A solenoid may also be used to operate the lever, but it must be in parallel with the circuit with 110 volts applied to it.

A tilting tank, Fig. 3-44, is another method for dispensing the detergent in the second wash. In the first wash both sides of the tank are filled with water which runs out of one side during the drain period between wash cycles. This will unbalance the cup, causing it to tip. The pivot pin must be straight to allow the tank to pivot freely when the door is upright. The screen should be cleaned periodically to permit the water to filter through it. If the screen is mounted across the top of the opening and it becomes clogged, the cup will remain unbalanced and will tip in the first cycle, releasing the detergent. A clogged screen positioned vertically in the tank will prevent the water from draining out of the small hole and keep the cup from tipping. This will reduce the wash action and the detergent will harden in the second cup during the dry cycle. If any of the detergent enters the last rinse cycle and remains on the glassware during the drying it may cause an etching effect on the glasses.

RINSE ADDITIVE DISPENSERS

Rinse additives have been developed within the past few years to prevent mineral spotting on the glasses. The additive breaks the surface tension of the water so it will run off in a sheet, eliminating the formation of droplets which dry and leave a deposit. The additive is usually stored in a tank and injected into the last rinse. Eight to ten drops are all that is needed for the six to ten quarts of water used in the rinse cycle. The dispenser may have a piston which fills a chamber and is triggered by a solenoid or heater element to release the solution. It may have ball checks that control the flow of the liquid. The liquid level should be checked occasionally and the dispenser refilled when necessary. A water leak may activate the solution and corrode the mechanism.

Another type of dispenser is energized by a bimetal strip in series with the motor or heater which raises a plunger to release the solution. Opening the door tilts the tank assembly attached to it; this fills a chamber which holds the correct amount of additive and later dispenses it when the plunger is raised. If the plunger is not firmly

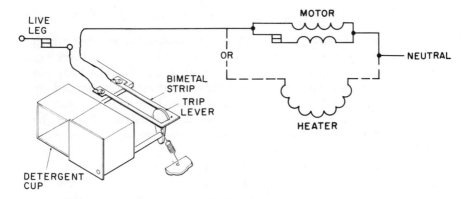

Fig. 3-43. Circuit for bimetal strip.

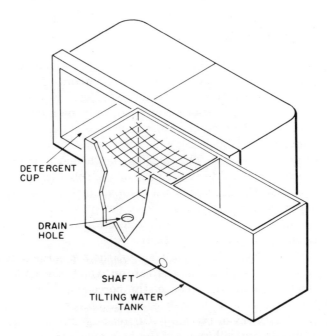

Fig. 3-44. Tilting tank.

seated, the additive in the well may run out after each door opening; an adjustment screw which sets the tension on the plunger must be tightened.

DISH RACKS

The dishwasher racks are made of plastic coated steel which cushions the dishes and prevents them from chipping and marking. Occasionally, the coating may wear down to the metal which will rust unless it is properly repaired with a coating recommended by the manufacturer or epoxy available at most hardware stores.

Dish racks are arranged in different ways, depending on the wash principle used. For example, an impeller model has a circular dish rack facing in the direction of the impeller rotation. A deflector may be used in the center of the rack to control the spray. If the rack is not positioned properly over the impeller or if the deflector is shifted, water may be directed toward a vent in the door and cause a leak. The use and care book for a dishwasher describes the way in which the particular dishwasher should be stacked.

DOOR ADJUSTMENTS

A door leak may be difficult to eliminate because of the difficulty in making proper adjustments. Some doors are adjusted at the bottom, either in and out or up and down, Fig. 3-45A, while others have a permanent hinge position and can only be adjusted at the top, Fig. 3-45B. Top loading machines, although not as susceptible to leaks as front loading models, may have a poor seal on the latch side if the hinges are set too low. The impeller or wash-arm action, which sprays the water upward from the base of the tub, directs the water toward the top of the door, which develops a leak if the corners are not seated properly. When the door is set down too far or in too tight at the bottom, it can have a wedging effect which tends to keep the top of the door from seating properly.

It is often difficult to see the leak at the top corners of a door because of the recessed lip on the tub on which the gasket seats and because of the glossy finish of the porcelain. These conditions can make it appear as if the leak is at the bottom of the door, requiring an adjustment of the hinge. Increasing the pressure on the gasket at the bottom will increase the leak which is at the top. On some models the gasket is omitted from the bottom of the door to increase the air circulation over the heater unit or to provide a vent during the drying cycle. It may be necessary to release the pressure at the hinge in order to tighten the latch adjustment at the top. However, the gasket may have taken a permanent set and

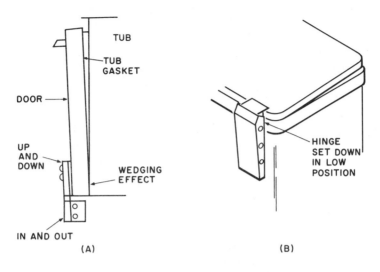

Fig. 3-45. Door adjustments.

require replacement. Door latches are adjusted by either loosening the mounting screws to reposition the latch or by adding or removing shims behind the catch. The adjustments should be made gradually to preserve the gasket and to allow for additional adjustment.

The door switch is activated by the latch mechanism or a bracket on the door. It is in series either with the entire circuit or that part which controls the water, timer motor, and wash motor. After checking the power supply at the terminal block or receptacle, check or jump the door switch. The connections are usually most accessible at the timer where a jumper could be placed or a test light secured to a ground terminal and the incoming or outgoing contacts checked. If the switch is fed directly from the terminal block it would be necessary to place a jumper from the main terminal to the single contact on the timer. Mounting screws may have loosened or the plunger on the door switch may have become worn from usage and will not make contact when the door is closed. If the switch plunger is accessible, set the timer at a mid-position in the wash cycle and, with the door partly closed, depress the plunger with a small screwdriver. If the motor starts this indicates that an adjustment or replacement of the switch is necessary. The contacts may also be in a condition where they open from the vibration of the machine. This would interrupt the cycle but the machine would restart when the latch is reopened and closed, making it difficult to locate the trouble. Occasionally, the contacts may fuse, which would allow the machine to run with the door open; a switch may also short circuit to the frame and blow the fuse. Usually, the switch is in the live side of the line and a new fuse will blow immediately after it is installed.

DRAIN SYSTEMS

Either of two types of drain systems, gravity or pump, are used in dishwashers, see Figs. 2-8 to 2-10.

Gravity Drain

The gravity drain must be located below the tub level and fed through a 1-1/2 inch tailpiece to a trap connected to the main waste line. The water is usually retained in the tub by a drain valve which is controlled through the timer and activated by a solenoid. The valve seat may be metal or rubber and controlled by an arm or a lever which holds the seat closed. If the lever is worn or the seat is not closing properly the water will run out during the wash cycle. The packing around the lever shaft may leak and appear to require replacement. Actually, a drainage problem in the main line causing the water to stay in the line too long can cause seepage at the packing which would not happen if the water drained out properly. If the drain line backs up and some grease from the sink coats the face of a flapper type valve seat, the valve may stick and prevent the water from draining. A solenoid may become magnetized and not release on every cycle, making it more difficult to diagnose the leak. However, if it does happen while the cycle is being checked, opening the door handle will open the circuit and the solenoid should release; otherwise replace the solenoid.

When a perforated ring or collar is used around a drain opening, Fig. 3-46, the water will filter through it. If the ring is fitted to the impeller it will keep the water back when the impeller is rotating.

The ring should be positioned as close to the tub as possible. If it is set too high or loosely fitted in the impeller, the water can drain out during the wash cycle. Using a piece of tape to build up a loose ring for a tighter fit may make it difficult to align the ring

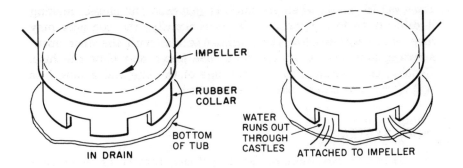

Fig. 3-46. Collar around drain opening.

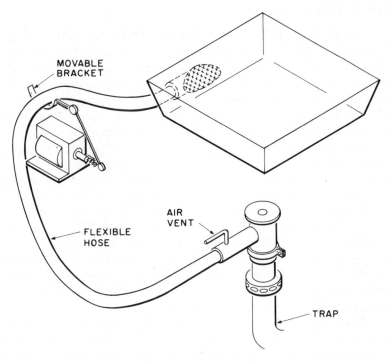

Fig. 3-47. Hose used as a gravity drain.

with the tub; a loud whistle may occur during the drying cycle when the water dries out between the ring and the tub.

A flexible hose may also be used in the drain system, Fig. 3-47. When the hose is raised above the water level by a bracket operated by a solenoid, the tub will not drain. This system is efficient unless there is poor drainage in the main line or the hose is twisted to meet the trap connection. Under these conditions, water will not run off properly and may siphon out in the next wash. A small air vent is installed in the hose to prevent this siphoning.

Pump Drain

The pump drain is probably the most popular and least expensive device for removing water from the tub. A common pump used for draining is self-contained and driven by a small shaded-pole motor, Fig. 3-48, or from the bottom of the main motor. A plastic or metal pump housing and a blade sealed with a stationery or rotary type spring seal around the shaft is coupled to the motor shaft. A rubber gasket is used to form a seal between both sections of the pump housing. It is important to tighten the screws evenly to assure a proper seal.

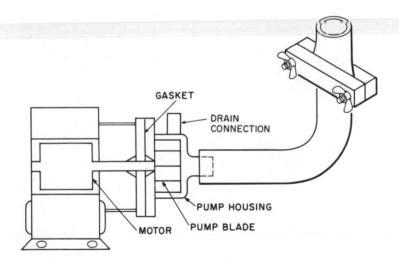

Fig. 3-48. Typical pump.

If the housing requires dismantling to check and remove whatever is binding the pump blade, it will be necessary to pinch off the section of drain hose between the tub and the pump to keep the water in the machine. With a clamp or a pair of wide jaw grip pliers, clamp the hose about midway. This will permit cleaning of the lower section and removal of any other particles that could jam the pump blade again.

The condition of the blade shaft should always be checked when a stationary type seal must be replaced. If the shaft is grooved the replacement seal will also become damaged. The proper alignment of the shaft and the seal is important to prevent leakage. A bent shaft or worn motor bearings must be corrected before the new seal is installed.

A drain valve may be used to control the flow to the drain pump when the pump is driven by the main motor instead of its own motor, Fig. 3-49. Addition of a bypass makes it possible to prevent the water level from exceeding a maximum height. If the pressure switch does not cut off the water valve, the excess water will overflow into the bypass and be pumped out during the wash cycle.

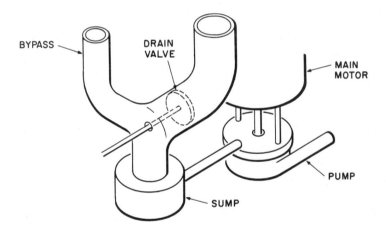

Fig. 3-49. Drain pump driven by main motor.

A worn valve seat will allow the water to enter the drain line during the wash cycle, gradually reducing the level in the tub, which will affect washing efficiency.

Another method of pump action incorporates the pump blade in the sump of the tub and reverses the motor for the pump cycle. The blade is designed to force the water downward and out through

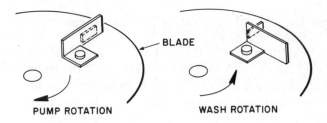

Fig. 3-50. Impeller swivel vanes.

a port in the drain casting and eliminate the valve in the drain line. The blade may be designed with fixed vanes or swivel vanes, Fig. 3-50, that move into position when the motor direction is in the direction of pump rotation. These swivel vanes must always move freely for an efficient pump out.

PLUMBING

A set of installation instructions accompanies each machine to help the plumber and electrician install the machine correctly. The instructions should be followed exactly to ensure a sufficient pump out for the time allotted in the drain cycle.

To prevent a drain off from the pump, because the line is installed below the level of the tub, a section of tubing is supplied (or must be purchased) as an accessory to ensure adequate installation. Even though these parts are available, there are many machines that cannot operate satisfactorily because of the limitations of installation. See also Fig. 2-10.

To eliminate the pulsating noise of the pump after it has emptied the machine, a check valve may be installed in the drain

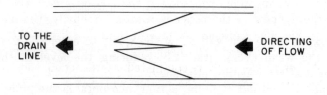

Fig. 3-51. Flapper valve installation.

line, Fig. 3-51. The valve may be a rubber coupling, but with a flap-per valve on the inside that must be positioned in the proper di-rection; otherwise, it will restrict the flow. If the drain tubing is inserted directly into the coupling, do not insert it far in or it may butt the valve and prevent it from opening properly.

DRYING (HEATERS)

After the wash and rinse cycles are over, most dishwashers are equipped with heaters to dry the dishes. Some models have a separate fan while others use the impeller to circulate the warm air. The heater may also be operated without fan circulation using the natural convection of warm air rising from the heater and passing over the dishes. If the heater is an open coil type, it would be lo-cated outside the tub area. The heated air would be channelled through a duct into the machine. The types of heaters are described in Chapter 2.

The heater may be used during the wash cycle to help main-tain the water temperature, or an additional element may be built into the heater to raise the temperature. Some models use a thermo-stat in the circuit to sense the temperature of the water coming into the machine and to energize the heater until the proper temperature is reached before the wash motor can start.

It may be necessary to check the heater, which could be burned out or disconnected because of a bad connection either inside or outside of the timer. There are several ways to check the heater circuit but a clamp-on ammeter or a test light is effective for quick and accurate diagnosis. Clamp the ammeter around one leg of the heater lead and advance the timer to the heat cycle; the amount of current (if any) being drawn is shown on the ammeter. If the cir-cuit is open (no current flow), the use of a voltmeter or test lamp at the heater connections will show if the heater is defective or the supply line is open. If the timer is rotated manually, the meter or test lamp will also indicate an intermittent condition due to ex-cessive wear in the timer. The heat might go on for a short period and go off before the cycle is completed, resulting in poor drying.

A loose connection at the heater will cause a voltage drop and will overheat the terminal. Sometimes the screw will jam in the connector and it becomes difficult to tighten or remove. Grip pliers should be used to hold the tab and prevent it from breaking

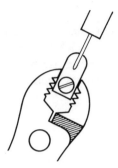

Fig. 3-52. Removing a connector that has a frozen screw.

off while trying to remove the screw, Fig. 3-52. If the screw slot becomes damaged the grip pliers may be used on the head of the screw while holding the tab with another pair of pliers. If the head breaks off it may still be possible to drill out the broken stud and save the heater. Before another screw is installed, use an 8-32 or 10-32 tap to clean out the thread.

Chapter 4

GARBAGE DISPOSERS

The garbage disposer has become a more popular appliance each year. In some areas, entire communities have been equipped with these units as part of their garbage disposal plan. The operation of the unit is quite simple and the unit may be installed in most sinks with a standard drain hook-up.

BASIC PRINCIPLES

A garbage disposer grinds or shreds garbage into small pieces which are then washed down the drain by water. Figure 4-1 shows the major components of a garbage disposer. The disposer consists of four major assemblies: (1) the mounting assembly, which attaches the disposer to the sink; (2) the hopper assembly, which stores the waste; (3) the flywheel and shredder assembly, which grinds the waste so that it can be disposed of through the house waste lines; and (4) the motor assembly, which powers the flywheel and shredder assembly.

Mounting Assembly

The mounting assembly consists of a sink flange, two mounting flanges, mounting bolts, and a lock ring. The sink flange is designed to most standard size sinks having a drain opening of about 3-1/2 inches. The shoulder of the flange seats around the

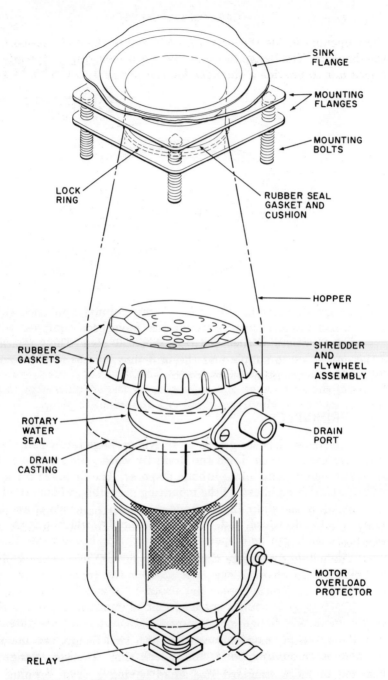

SINK
FLANGE

MOUNTING
FLANGES

MOUNTING
BOLTS

LOCK
RING

RUBBER SEAL
GASKET AND
CUSHION

HOPPER

RUBBER
GASKETS

SHREDDER
AND
FLYWHEEL
ASSEMBLY

ROTARY
WATER
SEAL

DRAIN
PORT

DRAIN
CASTING

MOTOR
OVERLOAD
PROTECTOR

RELAY

Fig. 4-1. Typical garbage disposer.

drain opening inside the sink, Fig. 4-2, and is the main support for the disposer. The flange has a groove for a lock ring or threads for a lock nut to provide a shoulder for the flange.

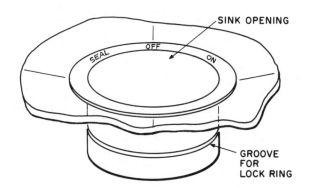

Fig. 4-2. Sink flange.

The mounting flange, Fig. 4-3, locks the disposer to the sink flange. A single or a double mounting flange may be used, depending on the particular disposer model. The flange is slipped over the throat of the sink flange and supported by the shoulder of the snap ring (or lock nut). Mounting studs lock the flange in place.

Hopper

The hopper, Fig. 4-4, which is the upper part of the disposer body, stores the waste that is to be ground by the shredder and flywheel assembly.

The amount of waste that the hopper can hold is determined not only by its physical size, but also by the capacity of the disposer motor. A typical hopper has about a 1-1/2-to-2-quart capacity.

The hopper is usually made of cast aluminum, and is attached to the mounting flange and fastened with nuts that are threaded into the studs. It is sealed and cushioned by a flanged rubber gasket that seats around the bottom opening of the sink flange, see Fig. 4-1. Another gasket seals it at the bottom where it meets the flywheel and shredder assembly.

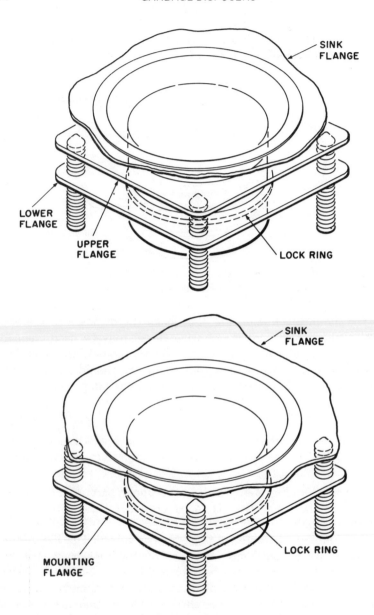

Fig. 4-3. Mounting flanges.

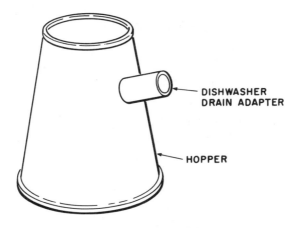

Fig. 4-4. Hopper assembly.

Hoppers on most disposers have knock-out embossments on the side. When the casting within the embossment is knocked out, an adapter line from the dishwasher drain can be connected to the disposer, eliminating the need for a separate drain line for a dishwasher.

Flywheel and Shredder

The flywheel and shredder assembly, see Fig. 4-1, grinds the waste contained in the hopper. The shredder, which is stationary, is the cutting edge of the assembly. The flywheel rotates at high speed to force the waste against the sharp edges of the shredder.

Shredder. The shredder is mounted at the base of the hopper. It is basically a hardened alloy ring with many castle-shaped openings, Fig. 4-5. These openings allow the shreddings to be flushed into the line drain after being cut up by the knife-like cutting edges around the inside of the ring.

Flywheel. The flywheel, Fig. 4-6, is mounted inside the shredder and is driven by the disposer motor. It is basically a steel disc with hammers on the top. The hammers force the waste against the shredder as the flywheel rotates. On some disposers the hammers are

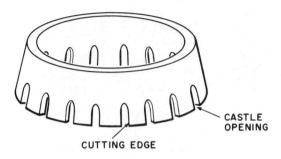

CASTLE
OPENING

CUTTING EDGE

Fig. 4-5. Shredder.

replaceable, while on others they are simply embossments on the flywheel and are not replaceable. The flywheel in many disposers has many small holes to allow water to drain off even if some underground waste blocks the shredder openings.

HAMMER

FLYWHEEL

Fig. 4-6. Flywheel.

A rotary carbon-faced seal, Fig. 4-7, is mounted under the flywheel to prevent water from entering the motor assembly. The lower half of this seal is spring loaded and has a carbon face that seats against the smooth center of the underside of the flywheel. Thus, the water is held in the drain section to the housing.

Should the carbon-faced seal start to leak for some reason, the water will be deflected into a small well by an umbrella type slinger. A small weep hole in the well will drain the water to the outside of the disposer to protect the motor assembly.

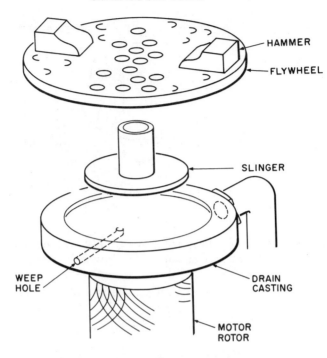

HAMMER

FLYWHEEL

SLINGER

WEEP
HOLE

DRAIN
CASTING

MOTOR
ROTOR

Fig. 4-7. Flywheel seal and weep hole.

Motor Assembly

Most disposers are driven by a conventional capacitor-start split-phase motor rated at about 1/3 to 1/2 horsepower, Fig. 4-8. The upper part of the motor housing may be used as the drain well that empties into the drain opening, see Fig. 4-1. The center of this housing is fitted with the upper motor bearing, which is protected from water leakage by the rotary water seal discussed previously. The lower part of this seal remains stationary, while the upper part rotates with the flywheel.

The rotor shaft may be threaded or splined into the flywheel. Sometimes, however, it is welded to the flywheel, and the rotor is slipped over the other end of the shaft and locked in place with a key and lock nut.

The stator windings are part of the lower housing, and consist of a start and run winding. The start winding is energized through

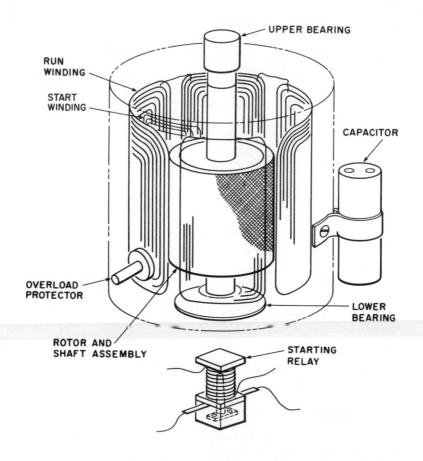

Fig. 4-8. Disposer split-phase motor.

a relay or a centrifugal switch, and only remains in the circuit mo-
mentarily until the rotor reaches approximately 2/3 running speed.
At this time the run winding takes over and continues to run the
motor at the normal operating speed of 1725 rpm.

To protect the stator windings from overheating due to a low
voltage condition or a jammed flywheel, a bimetalic overload protector is wired into the circuit. The overload protector is attached
to the winding to sense overheating and open the circuit when low
voltage or a jammed flywheel causes an excessively high current
to flow through the winding.

DISPOSER CONTROL SWITCHES

Control of the disposer is very simple. Unlike most appliances, there is only one control, the "on/off" switch, that concerns the user. One type of "on/off" switch is simply a toggle switch mounted in a conventional wall box. A second type of control, called a "switch top control" is an integral part of the disposer unit. One other control switch, called a "water flow control" switch is installed in the cold water line to the disposer, but is not activated by the user; it operates automatically when the sink faucet is turned on.

Wall Switch

The wall switch controlled disposer is operated by a conventional toggle switch, Fig. 4-9A, mounted in a wall box above the counter top. An alternate method is to mount a 3-inch gem box, Fig. 4-9B, on the side wall inside the sink cabinet. When it is mounted near the door opening on the cabinet, it is more accessible to the user.

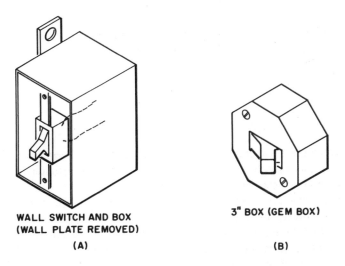

WALL SWITCH AND BOX
(WALL PLATE REMOVED)

(A)

3" BOX (GEM BOX)

(B)

Fig. 4-9. Disposer on/off switches.

A disposer unit using this type of control is a continuous feed unit. The sink flange is equipped with a sink stopper, Fig. 4-10A, to seal it when the sink is being filled with water. It seats on a rubber deflector, Fig. 4-10B, that acts as a baffle to prevent waste particles from being ejected when the motor is turned on.

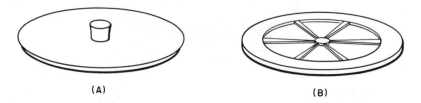

(A)   (B)

Fig. 4-10. Sink stopper (A) and rubber deflector (B) used with continuous feed disposers.

Switch Top Control

Switch top control is a more direct means of controlling the disposer. The "on/off" switch is mounted on the outer side of the hopper, and the switch button passes through a small opening in the hopper, Fig. 4-11. The switch is activated each time the switch top is rotated to the locked position. This also has an additional safety advantage because the top must be in place before the unit will operate.

Water Flow Control

A water flow control switch can be used in conjunction with either of the switches mentioned previously. The switch, Fig. 4-12, is installed in the cold water plumbing line, and operates automatically when the cold water faucet is turned on. As the water passes through the switch assembly, it raises the plunger in the direction of water flow. This will raise a lever attached to a shaft that turns an arm which will activate a microswitch.

The use of a water flow control switch provides two important advantages: (1) power cannot be applied to the disposer unless the cold water faucet is turned on, and (2) hot water cannot be used

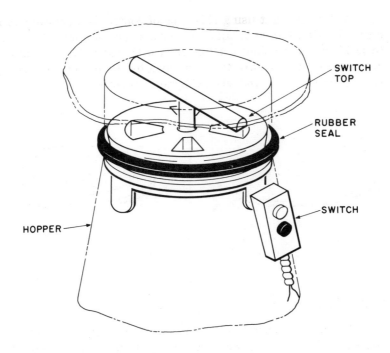

Fig. 4-11. Switch top control.

accidentally to flush the disposer. This is important because hot water causes grease to liquify and coat the plumbing lines, which may eventually cause a drain blockage.

INSTALLATION

Before installing a garbage disposer in a city or town having its own sewage system, check the local plumbing codes to make sure that garbage disposers are permissible. Garbage disposers can be installed in homes having septic tanks; the septic tank will simply have to be cleaned more frequently. If a house is in the planning stage, it is recommended that a 750-gallon tank, or larger, be installed.

Some people are under the impression that wastes from a disposer tend to block drains because some cities and towns prohibit their use. However, the prohibition is caused by the small

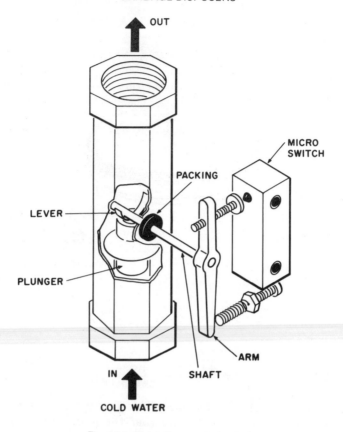

Fig. 4-12. Water flow control switch.

size and poor condition of some older sewage lines; but where lines
are adequate, the shredded particles will help clean the inner walls
of the sewage lines. In septic tanks, the ground garbage entering
the tank helps promote desired bacterial growth.

WATER SUPPLY

An adequate amount of cold water is required to ensure that
the shredded waste is flushed through the drain line. Too little
water will not carry away all the shreddings, which may reduce

drainage when the disposer is not in use. A flow of at least 2 gallons per minute is recommended.

Cold water is necessary because it helps to congeal grease which could coat the waste lines and eventually cause a blockage. Cold water is also utilized to cool the rotary seal mounted under the flywheel. This seal can be damaged without an adequate supply of cold water.

Water Flow Control Switch Adjustment

If a water flow control switch is used, it must be installed in the cold water line to the sink. Before adjusting the switch, remove the faucet aerator (if one is used) and clean it. Small particles that collect in the aerator reduce the water flow and may prevent the disposer from starting.

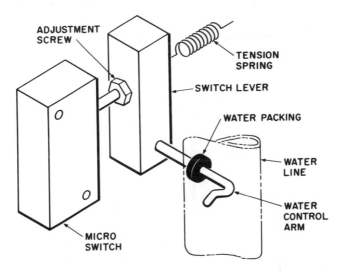

Fig. 4-13. Water flow control switch adjustment.

To adjust the switch, Fig. 4-13, loosen the lock nut on the adjustment screw and turn it slowly until the disposer starts with a normal flow of water and stops when the water is shut off.

Drain Connections

A garbage disposer can be installed in most sinks having a standard drain connection. Most disposers can be positioned to line up with the existing drain. With the crumb cup on the sink and the 1-1/2-inch tail piece leading to the trap removed, it is often possible to install the disposer by turning the lower half of the disposer to line up with the existing trap. If this is not possible because of the plumbing arrangement or because the lower half of the disposer does not rotate, then a swivel trap can be installed.

Mounting Unit to Sink

The mounting assembly secures the disposer to the underside of the sink. Attach the disposer to the sink as follows, see Fig. 4-14: (1) Apply an even amount of plumber's putty around the lip of the sink flange and then set the flange into the sink opening; (2) Slide the mounting flange over the throat of the sink flange and lock it in place with the lock ring; (3) Thread the studs through the mounting flange. The studs may be tightening directly against a cast

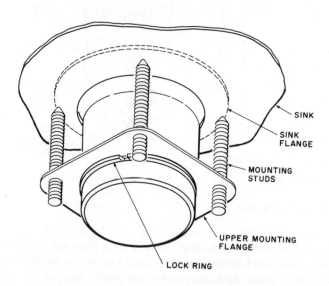

SINK

SINK FLANGE

MOUNTING STUDS

UPPER MOUNTING FLANGE

LOCK RING

Fig. 4-14. Attaching mounting assembly to sink.

iron sink or against a special collar placed against the underside of the sink. The collar prevents the screws from damaging the sink, and should always be used when the disposer is installed in a stainless steel sink. (4) Attach the hopper to the mounting flange and fasten it snugly by threading nuts onto the mounting studs.

In Fig. 4-14, a single mounting flange is shown. Some disposer models are provided with a double flange. When this is the case, attach the hopper to the lower flange.

If the disposer is a switch top control type, Fig. 4-11, the hopper will have to be properly aligned with the flange. Many disposers have a key slot in the sink flange body and an embossment on the mounting flange. The embossment on the mounting flange must line up with the slot on the sink flange before the mounting flange can be slipped over the throat of the sink flange, as in Fig. 4-15. This ensures that the "on/off" switch on the hopper and the switch top are properly aligned.

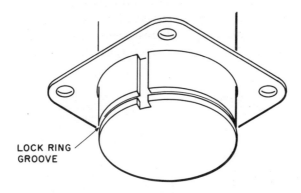

LOCK RING
GROOVE

Fig. 4-15. Sink flange alignment slot.

ELECTRICAL SUPPLY

Power Requirements

Garbage disposers require 115-volt, 60-cycle ac for operation. A separate line for the disposer, with a 15- or 20-amp fuse (or circuit breaker) should be installed. The fuse (or circuit breaker) should be of the time delay type to prevent it from opening on momentary overloads.

On/Off Control

On a continuous feed disposer, a toggle switch should be installed in the sink cabinet. This is the easiest installation in most cases. The a-c line from the switch is brought directly to the disposer motor, Fig. 4-16, or to the flow switch if one is used, Fig. 4-17.

On disposers with switch top control, the a-c line is connected directly to the disposer motor; no toggle switch is used. However, if a water control switch is used, the a-c line is brought through the switch to the motor, Fig. 4-18.

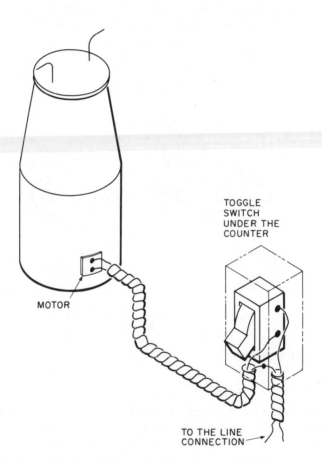

Fig. 4-16. On/Off toggle switch to control disposer.

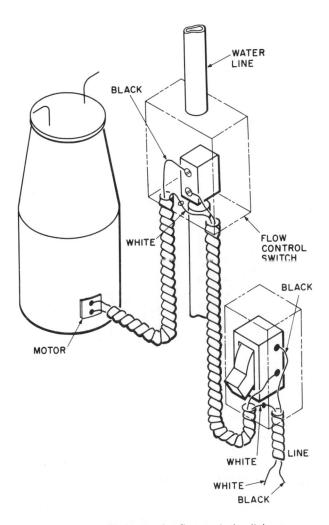

Fig. 4-17. Disposer using flow control switch.

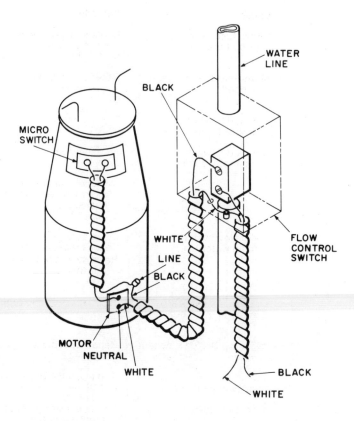

Fig. 4-18. Switch top control with water flow control switch.

MAINTENANCE AND TROUBLESHOOTING

There is not too much that can go wrong with a disposer. The major troubles are usually that it doesn't start, it makes noise, it leaks, or it does not drain properly. There are a number of possible reasons why any one of these troubles might occur. These are covered in the following troubleshooting chart. In a number of instances, the reader is referred to a specific page for further helpful information.

GARBAGE DISPOSER TROUBLESHOOTING CHART

Trouble	Cause	Remedy
No start	Blown fuse or tripped circuit breaker	Replace fuse or reset circuit breaker p. 105, 111
	Motor overload protector tripped	Reset or replace if necessary p. 111
	Jammed flywheel (motor hums)	Free flywheel p. 111
	Jammed flywheel (no motor hum)	Free flywheel p. 111
		Replace fuse or reset overload protector if necessary p. 105, 111
	Water flow switch (cold water line)	Clean aerator Check faucet washer Reset switch or reset microswitch p. 100, 111
	Open or shorted capacitor (some models)	Test capacitor and replace if necessary p. 97
	Twist top ON/OFF switch worn	Repair or replace p. 100, 112
	Twist top switch lever or switch plunger worn	Replace lever or switch p. 112
	Tight bearing	Replace bearing or complete assembly p. 113
	Defective centrifugal motor switch (some models)	Replace switch or adjust contact if necessary p. 97
	Inoperative relay (some models)	Check and replace p. 97
	Defective wall switch	Check and replace p. 106
Noisy	Piece of metal or other object jammed in shredder	Remove line fuse or turn off circuit breaker Remove or force object through shredder p. 111

Trouble	Cause	Remedy
Noisy (cont'd)	Loose or damaged blades or hammers on flywheel	Replace parts if available or replace complete flywheel p. 112
	Damaged seal (loud squeak)	Replace p. 113
	Loose flywheel	Lower disposer assembly and retighten or replace if worn p. 112
		Check seal p. 96
	Bearings worn	Check for side play
Water leak	Sink flange seal	Check and reseal flange p. 104
	Worn gasket at top of hopper	Check and replace p. 93
	Switch level packing leaking	Check packing or seal and replace p. 100, 112
	Worn seal (leak through weep hole)	Replace p. 97, 113
	Worn shredder gaskets	Check and replace p. 112
	Loose or worn slip nut gasket on drain	Replace washer and tighten p. 92
	Loose clean-out plug on trap	Retighten p. 92
Not grinding	Worn or stuck blades or hammers	Free blades or hammers Replace if necessary p. 112
	Worn or damaged shredder	Replace p. 112
Not draining	Blocked waste line	Remove and clean drain plug on trap Shake out line p. 92
Water draining and returning when disposer stops	Blocked drain or water going into vent	Clean out drain line p. 92

Motor Overload Protector

To protect the motor windings, particularly if the flywheel jams, a motor overload protector is used to break the circuit. The protector may be of the automatic type, which resets itself after a few minutes, or it may have a manual reset button. The reset button is mounted on the side of the motor housing. Before dismantling the unit, always check to make sure that the reset is not open (or the circuit fuse blown). If the overload protector continues to trip, a jammed flywheel is probably the source of the trouble.

Flywheel disposers are susceptible to occasional jamming due to foreign matter becoming mixed in with the garbage. When this happens, the foreign matter must either be removed or pushed through the shredder opening to allow the flywheel to move freely. The simplest household utensil that can be used to free the flywheel is a broom handle used as a lever against the hammer. Care must be taken because the wooden handle may snap. Special tools, such as "Z" bar or a handmade "T" handle, are more effective. After freeing the flywheel, the obstacle must be removed or the jamming will occur again. There are long tweezer type tongs available to clear the unit. Their use enables you to avoid having to dismantle the disposer to clean it. Where the obstacle is almost through the shredder, loosening the bolts slightly where the motor is fastened to the hopper may lower the flywheel sufficiently to allow the obstacle to pass through.

Water Flow Control Switch

It is important to maintain the correct switch adjustment so that enough water is supplied to flush the garbage out of the unit and through the drain. Before making any adjustment, remove the faucet aerator (if used) and clean it. Small particles of rust which collect at the screen will reduce the water flow and may prevent the disposer from starting.

A loose faucet washer screw may also act as a partial restrictor at the seat. If an adjustment is required, refer to the installation procedures given previously.

Switch Top Control

On disposers equipped with a control switch activated by the twist top, either the switch or the twist top itself may become defective. A packing nut or a rubber seal keeps water and food particles from entering the shaft bushing on the switch. This prevents binding and also eliminates corrosion in the switch linkage. Replace the packing or seal if there is evidence of leakage. A toggle switch may require an occasional adjustment to compensate for any wear in the linkage. Occasionally, the lever may become worn where it makes contact with the twist tip. Building the lever up with silver solder, or soldering a brass sleeve around the worn end of the lever, may restore the operation of the disposer until a replacement switch is obtained. After extensive use, the twist top may become worn to the point where it no longer operates the lever. If this occurs, replace the top.

Flywheel and Shredder Assembly

The most common troubles in this assembly are leaks and worn flywheel hammers. A leak may develop if the shredder ring is cracked or if the gaskets are worn or not seating properly. Either the shredder or gaskets can be replaced by separating the motor and the hopper. A screwdriver may be used to pry the shredder from the hopper housing. It is possible to remove the lower half of the disposer without removing the hopper, but it is probably easier to disconnect the entire unit at the mounting ring. In either case, it will be necessary to disconnect the tail piece to the trap. The flywheel hammers must be in good condition and move freely for good shredding action to take place. On some disposers, the worn hammers are replaceable; if they are not, a new flywheel has to be installed. When this is necessary, the flywheel must be removed from the rotor shaft. Some flywheels are fitted to the shaft using a spline or keyway and held in place with a locked bolt. When the bolt is replaced, it is important to use the proper sealant to prevent water from entering the motor compartment.

Some flywheels are threaded onto the shaft. If so, the rotor must be locked in a vise or held stationary and the flywheel hammer tapped with a rubber or plastic mallet in the direction of flywheel rotation. When the stator is removed from the rotor assembly, it is possible to support the rotor in a vise, using wood or leather between the jaws to avoid damaging it. Pliers or a wrench should not be used at the bottom end of the rotor shaft on a double-bearing motor assembly because the bearing surface may be damaged. Pliers or a wrench may be used if the motor has a single upper bearing and the shaft extension is not used.

Never plug the weep hole to eliminate a leak. With it plugged, the water will be directed into the motor compartment, which may result in damaged bearings or motor burnout.

Place a small amount of rubber cement around the outer edge of the bottom half of the seal. The rubber cement will act as a lubricant as the seal is being inserted, and will create a tight seal when it dries. Add a few drops of a high grade oil on the face of the seal to keep it properly lubricated. After replacing the seal, tighten the flywheel against the seal face to compress it.

Chapter 5

ELECTRIC RANGES
AND OVENS

Electric ranges and ovens are sometimes constructed as a single unit and at other times as two separate units. In some homes, you will find a range unit installed in the kitchen counter top and the oven installed separately in the kitchen wall. Both units have separate operating controls and are independent of one another. In other homes, the range and oven are constructed as one unit, with the oven mounted directly under, or above, the range. Some units even have two ovens, one below and one above the range. Regardless of the physical construction of the various ranges and ovens, their operating principles are essentially the same. It is important to understand these principles to repair any type of range and oven quickly.

BASIC OPERATING PRINCIPLES

Figure 5-1 shows the major components of a combination electric range and oven. The functions of each of these components are described below.

> *Connection (Terminal) Block:* At this point the main electrical supply is connected to the range.
> *Wire Harness:* The three wires connect the range switches to the terminal block.

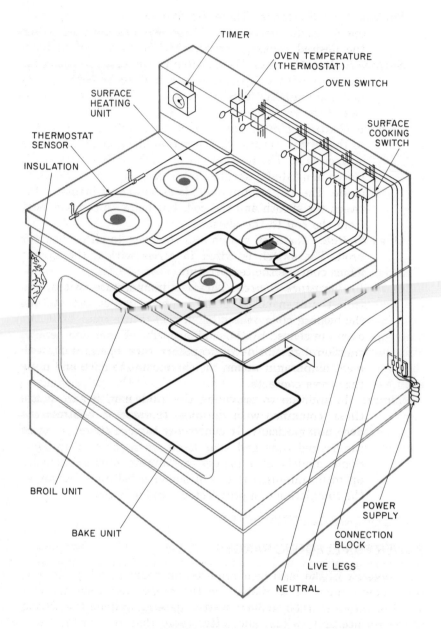

Fig. 5-1. Basic range construction.

Surface Unit Switches: There are four surface unit switches, one for each surface unit. These switches are used to set the desired cooking temperature of each surface unit.

Surface Switch Harness: Each switch is connected to the cooking units with a separate wiring harness to control each unit individually. This harness is connected to the load side of the switch.

Surface Heating Units: The surface unit coils are connected to the switch harness at a junction block attached to the surface unit. The heating coil of each surface unit is enclosed in a tubular sheath that is packed with magnesium oxide. The magnesium oxide is an electrical insulator between the sheath and the coil, but helps to conduct heat from the coil to the sheath.

Oven Switch: The oven switch sets the oven to either bake or broil. The switch is wired in series with a thermostat which controls the temperature.

Oven Temperature Switch: The oven temperature switch controls the oven heat through a set of contacts that turn the heating coils on and off. Temperature changes in the oven are sensed by the capillary tube. Expansion or contraction of gas inside the capillary tube operates the element diaphragm within the thermostat to open and close the heater contacts.

Timer: In addition to providing the kitchen with a clock, the timer, combined with the oven thermostat, controls the oven automatically. It can be set to turn the oven on at the desired time and shut it off when the food is done. Some models also include a "minute minder" usually up to 60 minutes) that will sound a bell or a buzzer at the completion of a preselected cooking interval.

POWER FOR ELECTRIC RANGES

Because of the high heat required for cooking, adequate electrical power must be available for the range. Each surface unit alone is rated at 1200 to 3000 watts, depending upon the setting of the surface unit switch; and, often more than one surface unit, or a surface unit and the oven, are used at the same time.

To provide adequate power most electric ranges operate off of a 220-volt, 60-cycle, single-phase line. Local electrical codes require a separate 3-wire line from the home fuse (or circuit breaker) panel to the range. No other electrical device may be connected to this line.

Although most electric ranges operate from a 220-volt, single-phase line, homes in some areas are supplied with 208 volts. A range designed to operate at 220 volts will not operate properly at 208 volts; one designed specifically to operate off of a 208-volt line should be installed. Conversely, a range designed for 208-volt operation should never be operated off of a 220-volt line because damage may result from the overvoltage.

RANGE SURFACE UNITS

The range surface unit consists of a group of surface heating elements, each with its own surface unit switch. The surface unit switch permits selection of the proper heat for cooking.

Heating Elements

A heating element may be of either a double-coil or a single-coil type, Fig. 5-2. In the double-coil type, which is the most common, heat is controlled by connecting the coils in parallel, in series, or by energizing only one of the coils, depending on the heat desired. At the same time, the voltage applied to the heating element is switched between 110 and 220 volts, depending on the heat required. When a double-coil element is used, either a rotary or a pushbutton type surface unit can be used, both of which provide heat in fixed increments, such as "low," "medium," and "high."

A single-coil heating element can be used if the surface unit switch is designed to control the heat thermostatically. The most common type of thermostatically controlled switch is the infinite heat switch, which provides continuously variable rather than fixed heat selection; it can also be used on a double-coil element if a jumper is placed across the two single elements.

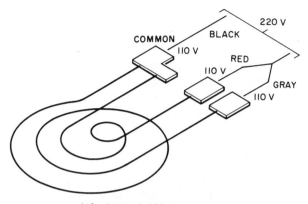

(A) DOUBLE COIL
INNER AND OUTER COILS

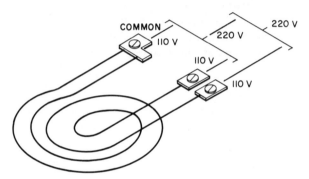

(B) INTERWOUND DOUBLE COIL

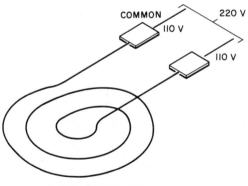

(C) SINGLE COIL

Fig. 5-2. Range heating elements.

Surface Unit Switches

The basic circuit of a surface unit switch is shown in Fig. 5-3. Keep in mind that most heating elements consist of two separate coils. The selection of high, medium, or low heat is controlled by switching the coils in and out of the circuit, depending on the heat desired, see Fig. 5-4. For example, in the high heat position, both coil 1 and coil 2 are energized by 220 volts. In the medium heat position, only coil 1 is energized (by 220 volts). In the low-heat position, coils 1 and 2 are connected in series and energized by 110 volts. Not only is the energizing voltage lower, but with the two coils connected in series, the resistance of the circuit is increased, cutting the current (and, therefore, the heat) even further.

The two most common switches used on most electric ranges are shown in Fig. 5-5. The rotary type makes its contact selections by a round can. The pushbutton type slides a bar which has protrusions that press against the moving contacts.

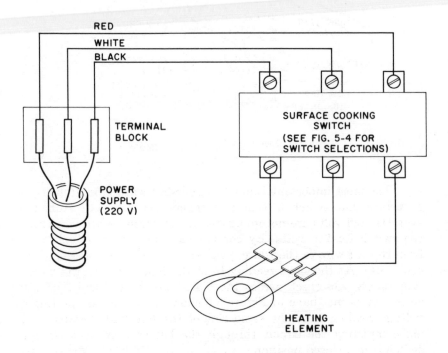

Fig. 5-3. Surface switch control circuit.

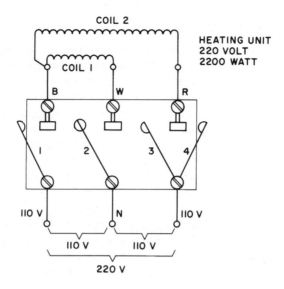

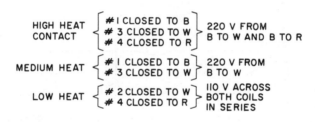

Fig. 5-4. Operation of surface unit switches.

The most important thing to remember when troubleshooting a surface unit switch is that you cannot get 220 volts from the switch if 220 volts are not going into it. Be careful when you check the switch for 220 volts. By connecting a test light or voltmeter from the red wire to the white, you can get a 110-volt reading. You could also get the same reading from the black wire to the white. But, unless you check across the black and the red and read 220 volts, you do not have it. Figure 5-6 shows why. You are getting a voltage reading from the same side of the line (red) because you are completing the circuit through the heater element when the switch is in a closed position.

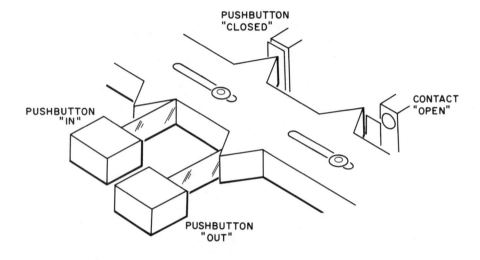

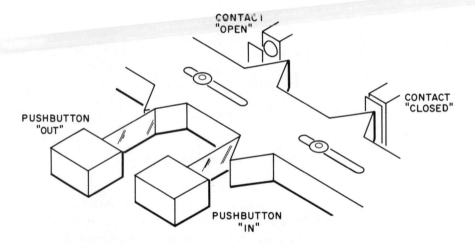

Fig. 5-5A. Pushbutton surface switch construction.

The infinite heat switch, Fig. 5-7, is a more recent development for controlling the heat at the surface units. Control is accomplished by passing the current through a bimetal strip which, in turn, opens and closes a set of contacts to maintain the heat at the desired setting. The switch can be set at any position from high to

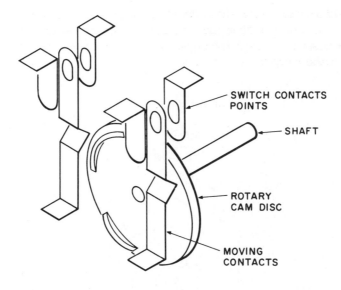

Fig. 5-5B. Rotary surface switch construction.

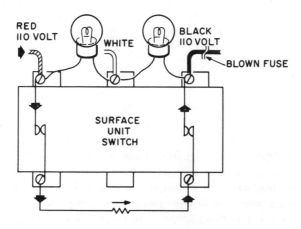

Fig. 5-6. Checking the surface unit switch.

low, and actually turns the complete unit on and off instead of eliminating a coil or reducing the voltage to reduce the heat. When the switch is set on "high," the coils are on continuously as they would be with the conventional switch.

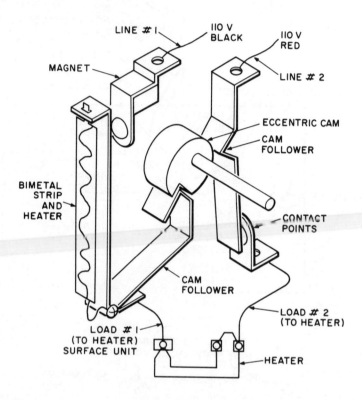

Fig. 5-7. Infinite heat switch.

Another method that is even more sensitive than the infinite heat switch is shown in Fig. 5-8. It has a sensing device in the center of the surface unit to actually measure the temperature at the pan. The sensor, which is a temperature sensitive resistor, controls the current through a hot-wire relay to operate the relay contacts and to maintain the desired temperature. As the temperature increases, so does the resistance in the sensor, opening the main circuit in the switch. As the temperature decreases, the sensor resistance drops, closing the switch circuit.

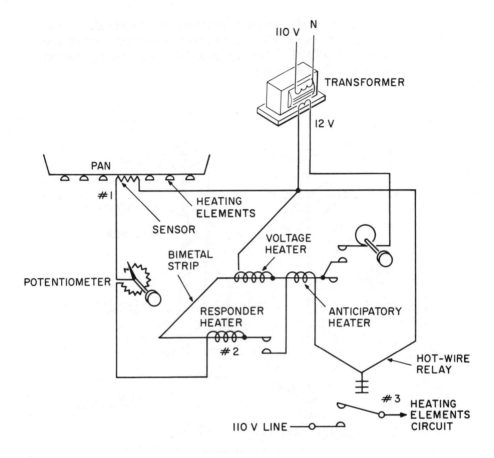

Fig. 5-8. Surface unit sensor switch.

When the sensor is cool its resistance is low, energizing the responder heater which moves the bimetal down to close contacts #2. This energizes the hot-wire relay, closing contacts #3 to complete the heating elements circuit. When the sensor temperature reaches the desired setting, contacts #2 of the responder heater open, deenergizing the hot-wire relay. The anticipatory heater, which cycles with the hot-wire relay, moves the bimetal arm in the opposite direction to open contacts #2 just before the sensor temperature is reached, providing a more accurate control. For a sensing unit to operate properly, the sensor must make contact with the bottom of the pan. If the bottom does not set properly on the unit because of its shape or because it is dented, it can affect the temperature control. If the sensor, which is spring loaded below the heating element for firm contact, becomes jammed in a low position, it will also affect the temperature setting.

OVENS

The oven unit, though it contains more components than the range unit, is still not complex. Unlike surface units, ovens contain features for automatically measuring cooking temperature, for automatically controlling the time that food is cooked, and even for self-cleaning. The components and circuits used to implement these features will now be discussed.

Heating Elements

An oven contains two, and sometimes three, heating elements. The two elements in all ovens are the bake element and the broil element. The bake element is mounted at the bottom of the oven and the broil element at the top, see Fig. 5-1. Broil or bake operation is set by the oven switch on the range. Both elements operate at 220 volts and require about 3000 watts.

Some ovens contain a third element, called a mullion heater. This heater, which cannot be seen, is mounted around the front of the oven, near the door. Its primary purpose is to provide additional heat at the front of the oven necessary for self-cleaning. The mullion heater operates at 110 volts and requires about 750 watts.

Temperature Controls

The proper oven temperature can be set by the oven temperature control (thermostat). The temperature control circuit is relatively simple; it consists essentially of the oven temperature control in series with the oven selector switch. The thermostat senses the temperature in the oven, and opens or closes the circuit to the oven heating elements to maintain relatively constant oven temperature.

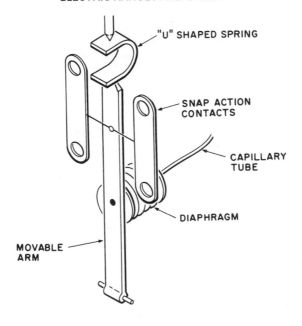

Fig. 5-9. Hydraulic thermostat.

A typical thermostat is shown in Fig. 5-9. If the temperature in the oven increases, gas in the capillary tube expands and exerts pressure against the diaphragm. The diaphragm, in turn, presses against the movable arm on the switch, opening the circuit to the heating element. The opposite occurs when the oven temperature drops below the setting of the oven temperature control. As the temperature drops, the gas in the capillary tube contracts, releasing the pressure on the diaphragm. This permits the switch contacts to close again. The thermostat must be wired in series with' the oven selector switch to open and close the switch circuit while the switch contacts remain closed, Fig. 5-10.

Electric Meat Thermometers

To roast meats accurately, insert an ordinary meat thermometer into the roast to a depth of 3 inches or so. How well done the meat is can be measured by the temperature inside the roast. How-

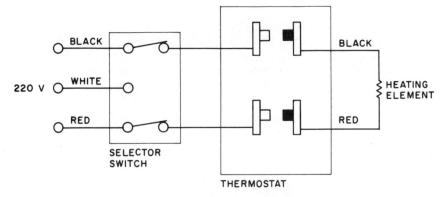

Fig. 5-10. Thermostat circuit.

ever, an occasional check is required and the meat may be overdone if it is forgotten. Many electric ranges are provided with an electric meat thermometer. The thermometer is basically a thermistor (a temperature sensitive resistor) contained in a probe. The probe is inserted in the meat and the temperature is measured through an electrical circuit which energizes a buzzer to signal that the meat is done to a preset temperature, Fig. 5-11.

The low-voltage circuit, which is stepped down from 110 volts by a transformer, passes through a bimetal strip to control the indicator and buzzer. The buzzer is controlled by the amount of current passing through the variable resistor in the meat probe. As the temperature increases, the current flow increases through the thermistor to warp the bimetal, raising the indicator pointer and energizing the buzzer when the meat temperature reaches that set by the manual pointer. The thermometer circuit is energized when the meat probe jack is inserted into the probe receptacle inside the oven.

Automatic Timer

To increase the versatility of the electric range, an automatic timer is included in many ranges. When a set of contacts are placed in series with the thermostat, Fig. 5-12, and operated automatically by a set of cams controlled by the clock, it is possible to turn the oven on and off. On some ranges, it is possible to control a receptacle or a surface element in the same way.

On some ranges the timer is bypassed except when it is in use; on others it is not. If the timer is always in the circuit, it may have a manual position that must be reset each time it is used. If this

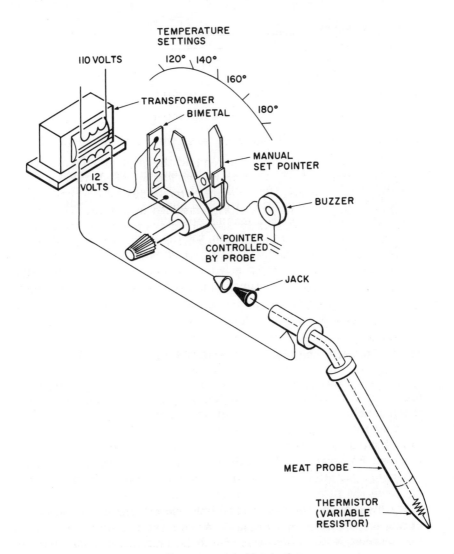

Fig. 5-11. Roast control mechanism.

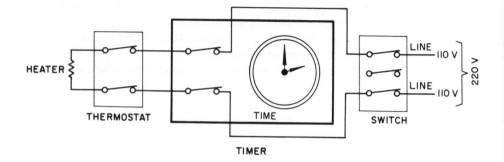

Fig. 5-12. Timer circuit.

is not done the oven will not go on, even though the oven switch is on "bake."

By adding another contact in the switch and a "time bake" position on the switch knob, Fig. 5-13A, the timer contacts are by-passed in the "bake" position, which eliminates the possibility of an open circuit when there is a manual position.

In addition to timing the oven, the timer can also be connected to the 110-volt receptacle to control an appliance such as a coffee maker, Fig. 5-13B. The receptacle may be controlled through a separate set of contacts in the timer on either the live or neutral side of the line.

MAINTENANCE AND TROUBLESHOOTING

It may seem difficult to service an electric range because of the number of interlaced wires that appear on the control panel. Often much time is spent in replacing unnecessary parts before locating the trouble or even failing to make the repair. Yet each branch circuit in the range is a simple circuit if we separate it from the rest. For example, if a surface unit is out, check to see if the others are operating correctly. If they are, this means the main supply from the fuse (or circuit breaker) panel is complete; otherwise everything on the range would be out. However, if all of the units only get warm, this would indicate an open fuse on one side of the line because in the lower heat-switch positions, the units operate on 110 volts across the 220-volt units. This will account for the

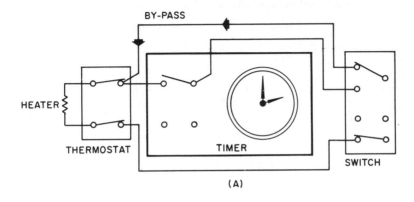

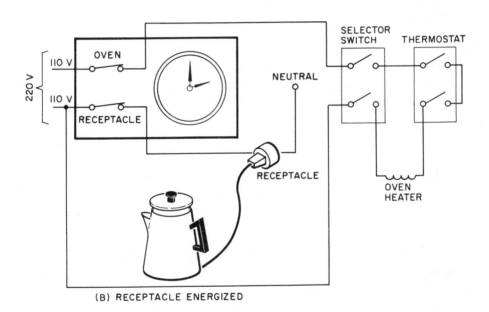

Fig. 5-13. Timer circuit switching arrangement.

low heat even when part of the circuit is open. Therefore, be sure to analyze the trouble before replacing any parts.

The following troubleshooting chart lists many of the common troubles that may occur in electric ranges. In many cases, the reader is referred to a specific page for additional information.

ELECTRIC RANGE TROUBLESHOOTING CHART

Trouble	Cause	Remedy
No heat (surface units and oven)	Blown fuse or tripped circuit breaker	Replace fuse or reset circuit breaker
No heat on high setting (surface units heat on low)	Blown fuse or tripped circuit breaker	Replace fuse or reset circuit breaker
One surface unit not heating	Broken connection	Replace p. 116
	Open heating unit	Replace p. 116
	Inoperative switch	Repair p. 120
Surface unit heating on only one unit	Broken connection	Repair p. 116, 120
	Open coil	Replace p. 116, 118
Oven units not heating	Open heating coils	Repair or replace p. 116
	Broken connection at unit	Repair p. 116
	Inoperative thermostat	Replace and check calibration p. 127
	Inoperative oven switch	Replace p. 121
Oven too hot	Thermostat out of calibration	Recalibrate Replace if necessary p. 127
No heat on TIME BAKE	Timer clock not operating	Repair broken connection or replace clock p. 128
	Timer clock high-voltage contacts inoperative	Replace p. 128
Uneven cooking (fry pan)	Range not level Surface unit not level	Level range, adjust legs Adjust unit or replace
Uneven baking	Range not level	Level range Align or replace racks
Clock accessory outlet and oven light not operating	Blown fuse or tripped circuit breaker in range	Replace fuse or reset circuit breaker (Check use and care book for location)

SELF-CLEANING OVENS

Probably the most outstanding feature in cooking equipment today is the self-cleaning oven. The most popular and widely used method works on the principle of oxidation without burning, known as "pyrolysis." Pyrolysis is accomplished by raising the temperature of the dirt above its flash point, which would normally cause ignition and burning, and limiting the oxygen supply by means of a very tight door seal.

In the cleaning process, which takes place at about 700-900° F, the moisture is removed first and the remaining hydrocarbons are then broken down into smoke and gases. Elimination of the smoke as it leaves the oven through the vent in the top is accomplished by a small heater set under a platinum-coated wire screen. The heated screen acts as a catalyst to complete the chemical reaction. The resultant carbon dioxide and water are then passed off into the air in small amounts that are harmless and not very noticeable.

Construction and Controls

The oven itself appears similar to the conventional oven except that there is a latch handle above the door that locks it during the cleaning process. Since additional heat is required for the self-cleaning process, it may seem that the heating units would have to be large and the operating costs very high. Actually, the broil and bake units are the same size as those for a standard oven, but the cost of operating them is less because they are supplied with 110 volts, which reduces the wattage to 1/4 of the rated wattage at 220 volts. For example, a unit rated at 3000 watts at 220 volts would reduce to 750 watts at 110 volts. In addition to the bake and broil units, some manufacturers incorporate a mullion heater around the outer edge of the oven opening to compensate for the heat loss at the door seal as shown in Fig. 5-14. The addition of the mullion heater increases the area of the oven that is cleaned right to the door opening. The mullion heater is also operated on 110 volts, which increases the total wattage by approximately 750 watts. The wattage for the three units totals 2250 watts, which is less than the wattage of the bake unit during the normal cycle. Depending on the amount of soil to be removed, the average cleaning time would be 2 to 2-1/2 hours, which keeps the cost reasonably low.

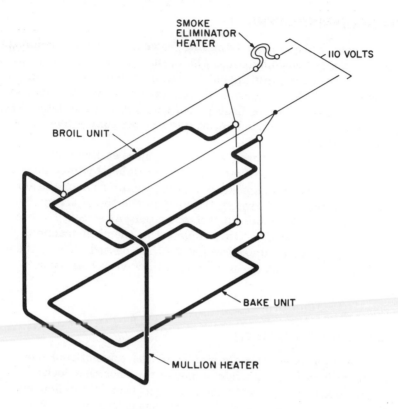

Fig. 5-14. Power circuit during self-cleaning.

A fiberglass gasket seals the door and withstands the heat that is maintained for self-cleaning. The smoke control system consists of a small heater which is wired in series with the oven units and positioned in the oven vent system. To maintain the proper temperatures on the outer enclosures of the free-standing ranges and wall oven, substantial insulation is needed. In addition, a circulation fan provides air movement to keep the unit as cool as possible. The fan is energized by a temperature limit switch that activates it when the door is locked and the heat is up to the cleaning temperatures.

Self-Cleaning Temperature Controls. In electric ovens without the self-cleaning feature, the thermostat is generally the hydraulic type that operates on the expansion and contraction of a gas enclosed in a long tube positioned inside the oven. As the gas expands and contracts from the heat contained in the oven, a small bellows or diaphragm activates a set of electrical contacts within the thermostat. Since there is a large variation in the temperatures that occur for the bake, broil and self-clean cycles, an electrically operated thermostat is used rather than a hydraulic one. With a variable resistor that changes its resistance value with temperature, the same control device can be used for both the high- and the low-temperature ranges.

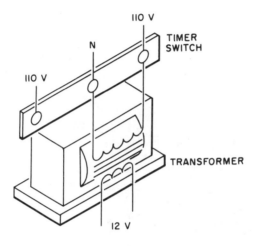

Fig. 5-15. Twelve-volt circuit for oven sensor control.

A step-down transformer reduces the 110-volt supply from one side of the 220-volt line to the 12 volts which operate the control. During the bake cycle the oven switch energizes the 12-volt circuit to operate the oven sensor circuit that controls the hot wire relay, as shown in Fig. 5-15.

Responder Type Control Circuits. The responder control circuit, Fig. 5-16, operates on low voltage and controls the oven temperature in the bake, broil, and cleaning cycles. A step-down transformer reduces the voltage from 110 volts to approximately 12 volts. The oven sensing device is a variable resistor which is enclosed in a metal tube and positioned in the oven. The oven sensor is in series with a potentiometer that manually sets the oven temperature. Both the sensor and potentiometer are in series with a responder, which opens and closes as temperature changes. As the temperature goes up, the resistance increases, reducing the current flow and opening the responder contacts. Then as the oven cools, the resistance of the sensor decreases, allowing more current to flow, closing the responder contacts. This activates the hot-wire relay and closes the high-voltage contacts to heat the oven.

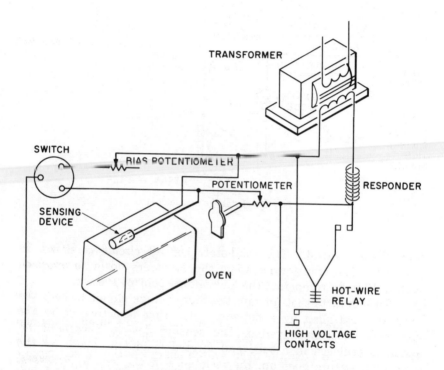

Fig. 5-16. Self-cleaning responder control.

For self-cleaning, the oven latch handle is placed in the locked position, thereby closing the switch. This puts the bias potentiometer in parallel with the sensor. The current flow increases and the value of the sensing circuit changes to keep the heating units on longer for the cleaning cycle.

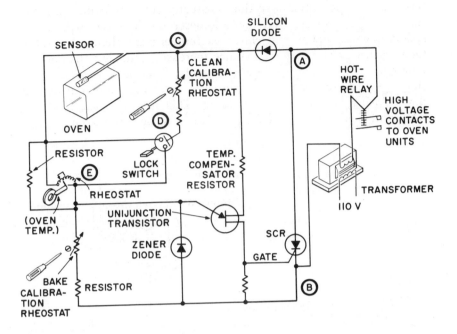

Fig. 5-17. Solid-state self-clean control.

Solid-State Control. The solid-state control circuit is shown in Fig. 5-17. The transformer steps the 110 volts down to approximately 12 volts to operate the low-voltage control system.

The contacts that operate the high-voltage circuit to heat the oven are controlled by a hot-wire relay that is activated by the solid-state electronic system. The sensing device, a variable resistor, is enclosed in a tube and located inside the oven. As the oven temperature goes up, the resistance of the sensor increases, reducing the current flow in the circuit, which opens the hot-wire relay controlling the high-voltage contacts.

A rheostat in series with the sensor controls the temperature required for the bake and broil cycles. A unijunction transistor is a d-c device, which allows a current to flow in one direction only. To supply the transistor, a converter or rectifier is necessary to change the a-c supply voltage at the secondary side of the transformer to the d-c voltage required by the transistor. This is accomplished by the silicon diode in the circuit.

The unijunction transistor trips a switch to complete the circuit to the hot-wire relay. However, the switch that we are referring to is not a manually controlled device, but an electronically controlled one; referred to as an SCR (silicon controlled rectifier).

When the voltage reaches a certain value at the SCR gate, it triggers the SCR and closes the circuit to the hot-wire relay. When the hot-wire relay is energized, its high-voltage contacts close to operate the oven. To complete the control circuit and bring the oven heat up to its cleaning temperature, another circuit must be added to change the range of the cycling control.

The lock switch is activated with a manual lever which locks the oven for the cleaning cycle. This puts another set of resistors in parallel with the oven sensor and locks out the temperature control rheostat. This changes the value of the sensing circuit and allows the oven temperature to reach the higher cleaning range.

Troubleshooting

Troubles that may appear in a self-cleaning oven are given in the following chart. These troubles are in addition to those in the table given earlier in this chapter for conventional ovens. Be sure to check that table also when troubleshooting a self-cleaning oven.

SELF-CLEANING OVEN TROUBLESHOOTING CHART

Trouble	Cause	Remedy
No heat in oven (surface units not heating)	Blown fuse	Replace
No heat in oven (surface units heating)	Oven sensor open	Replace p. 135
	Temperature control-high-voltage contacts open	Check and replace p. 136
	Temperature control-defective hot-wire relay	Check and replace p. 136
	Temperature control-defective response contacts	Check and replace p. 135
	Temperature control-electronic circuit open	Check and replace p. 137

	Heater for smoke elimination open	Replace p. 134
	Open transformer	Replace p. 135
	Inoperative oven switch	Repair broiler wires or replace switch
Not cleaning	Not latched	Customer did not close latch, or Defective oven switch on latch mechanism Repair or replace p. 138
	Open bias circuit in control	Replace p. 135, 140
Partial cleaning	Timer not set for required time	Reset and try again
	Bias circuit resistance set too high	Recalibrate p. 135
	All heating units not operating	Repair wiring if necessary, or Replace heaters p. 126
	Defective timer	Repair or replace p. 128

Self-Clean Checking Procedure

When checking out the operation of a self-cleaning oven, be sure to understand the principle of the control device. This is the major difference between a conventional electric range and the self-cleaning range. The control is a low-voltage device used to control both the bake and the clean temperatures. If the control is not supplied with the necessary voltage to operate it, the high-voltage contacts used to energize the heating coils will not operate. Therefore, the first step, if the oven does not work, is to check the transformer, which reduces the line voltage of approximately 110 volts to the required 12 volts supplying the control circuit. By checking the low-voltage side first, we can tell if the transformer is working. If it is not, then check the input or the primary voltage which is controlled by the oven switch. Check as follows:

1. With the bake switch on, remove the transformer leads at "A" and "B", Fig. 5-17. The voltage across "A" and

"B" should be approximately 12 volts ac. Reconnect leads to "A" and "B."

2. With the lock switch in the unlocked position, remove the leads from "C" and "E." Measure the resistance through the sensor and through the oven temperature rheostat. Rotate knob from "warm" to "broil;" resistance will vary 15 to 45 ohms.

3. With the switch turned to "bake," check the voltage across "A" and "B." Both an a-c and a d-c voltage should be present. If an a-c voltage is present, but no d-c voltage is present, the solid-state circuit is not operating properly.

4. With the door unlatched, check the resistance from "C" to "D." If an open circuit is measured, the bias circuit for the self-cleaning cycle is inoperative, and self-cleaning cannot take place.

5. With the switch locked (closed), remove the sensor lead at "C." From the sensor lead "C" to "D" on the board, we should get the same sensor resistance reading at each connection on the switch. This indicates a good switch.

Adjustment of Clean Calibration Rheostat

Using a test resistor with the correct ohm rating (see manufacturer's specifications), disconnect the sensor lead at "C" on the printed circuit board. Connect the test resistor between "C" and "E." With the oven and the latch in the clean cycle, rotate the clean calibration rheostat until the oven cycling light is the brightest. Slowly turn in opposite direction until light dims (an audible click from the relay should also take place). Then, by slowly returning to the brightest light, the clean cycle is calibrated.

INDEX